EARTH'S OUTER ATMOSPHERE

BORDERING SPACE

GREGORY L. VOGT, Ed.D.

TWENTY-FIRST CENTURY BOOKS • MINNEAPOLIS

Twenty-First Century Books
A division of Lerner Publishing Group
241 First Avenue North
Minneapolis, MN 55401 U.S.A.

Website address: www.lernerbooks.com

Library of Congress Cataloging-in-Publication Data

Vogt, Gregory.
Earth's outer atmosphere : bordering space / by Gregory L. Vogt.
p. cm. — (Earth's spheres)
Includes bibliographical references and index.
ISBN-13: 978-0-7613-2842-1 (lib. bdg. : alk. paper)
ISBN-10: 0-7613-2842-4 (lib. bdg. : alk. paper)
1. Atmosphere, Upper—Popular works. 2. Atmosphere—Popular works.
I. Title.
QC879.V64 2007
551.51'4—dc22 2006019426

Manufactured in the United States of America
1 2 3 4 5 6 - DP - 12 11 10 09 08 07

CONTENTS

INTRODUCTION

SWIRLING LIGHTS

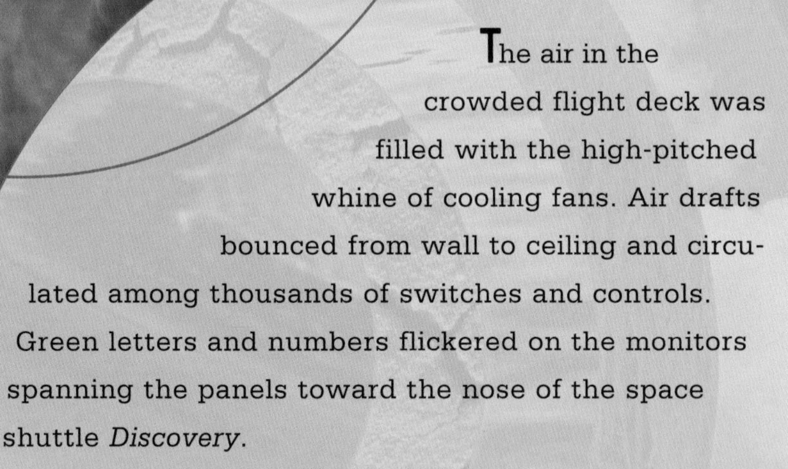

The air in the crowded flight deck was filled with the high-pitched whine of cooling fans. Air drafts bounced from wall to ceiling and circulated among thousands of switches and controls. Green letters and numbers flickered on the monitors spanning the panels toward the nose of the space shuttle *Discovery*.

The commander and pilot of the STS-39 mission were busy monitoring systems and reviewing orbital maneuvers they would have to perform when they returned to Earth. Oceans, islands, mountains, forests, and deserts were all flashing by at a speed of 5 miles (8 kilometers) per second. Earth's Southern Hemisphere was 160 miles (260 km) beneath them.

Toward the back of the flight deck, an astronaut was staring out the two windows facing the payload bay. Through these windows, the crew can operate payloads mounted in the bay and manipulate the 50-foot-long (15 meters) robot arm. But it was not the arm that had captured the attention of the astronaut. *Discovery* had passed into the dark nightside of Earth, but the sky was still aglow.

Above most of Earth's atmosphere and stretched out in front of and below *Discovery*'s orbit were swirls and rays of intense greenish light hanging against the nearly pitch-black background. These ghostly displays are called auroras, or northern and southern lights, depending upon which

The crew of the Space Shuttle STS-39 mission took this picture of the aurora australis, or southern lights, while orbiting above Earth's Southern Hemisphere.

hemisphere of Earth you are in when you see them. The light show slowly moved and danced, dimmed and flared. It was an astounding ghostlike display. The astronauts soon powered up a camera and took photographs.

In the foreground of the photos was *Discovery*'s vertical stabilizer, or tail fin. *Discovery* was tilted to its port side so that its left wingtip was pointed down at Earth and the stabilizer was parallel to Earth's surface. A white glow silhouetted the rounded orbital maneuvering system, rocket engine pods flanking the sides of the stabilizer. The glow came from faint traces of oxygen gas that collided with the pods as *Discovery* streaked across the sky. The impacts energized the widely spaced oxygen atoms and caused them to glow momentarily before they returned to their normal quiet state.

Other crew members took turns at the windows to enjoy the aurora light show. The display soon faded because *Discovery* passed from the nightside to the dayside of Earth, where sunlight overpowered the faint colors. It would be another forty-five minutes before they could round Earth to the nightside and see the lights again.

For many astronauts, the best part of spaceflight is looking at Earth. Things Earth dwellers take for granted, such as sunsets and sunrises and land and water, take on a new beauty when seen from 160 miles (260 km) away. Earth is a magnificent planet of rock, metal, water, air, and living things. Yet Earth is a relatively small planet when compared to some of the other planets in our solar

A full-hemisphere view of Earth taken in 1997. This image was created by using data from three satellites and shows a hurricane approaching Baja California in Mexico.

system, such as Jupiter or Saturn. Jupiter is eleven times as wide as Earth. Saturn's outermost rings could easily surround Earth and the Moon in its orbit. What sets Earth apart from the other planets is its incredible structure and diversity. Earth is a planet of major and minor spheres, some visible and some invisible.

Deep inside Earth where no one can see, Earth has a spherical core. The core is two-layered, with solid iron and nickel metal at its center, surrounded by a thick molten layer of the same two metals. Above the core is a spherical layer called the mantle. It is made of super-hot rock that slowly deforms and rises and falls with heat currents.

Covering the mantle is a relatively thin, hard, rocky layer called the lithosphere. The lithosphere is the surface, or crust, of Earth that we walk on. It is made up of all the continents with their mountains, valleys, and

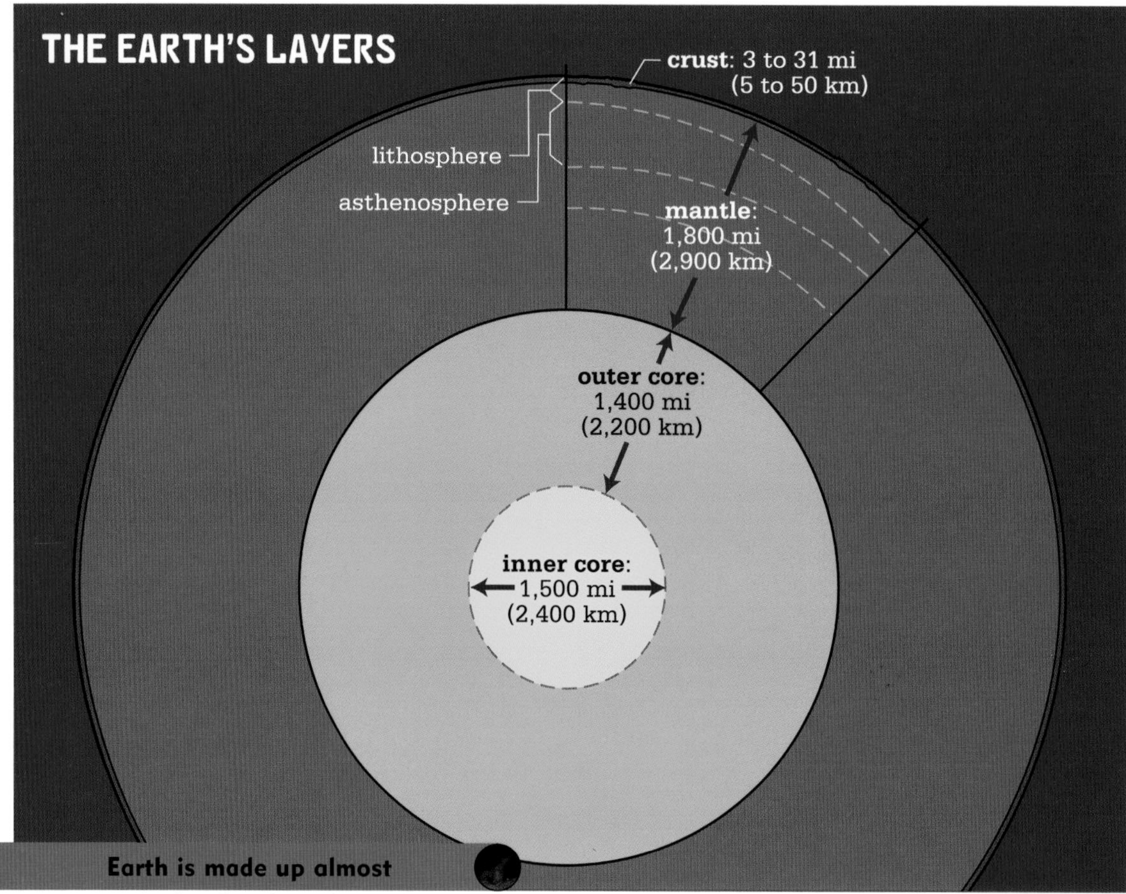

Earth is made up almost entirely of rock and metal. Its air, water, and life zones (atmosphere, hydrosphere, and biosphere) are far too thin to appear in this cross section diagram showing Earth's inner layers.

plains, as well as the ocean basins. Water is in the lithosphere's small depressions and wide basins and within the spaces between mineral grains in the soil and sub-surface rock. Water surrounds approximately three-quarters of the surface of the crust with great oceans. All the water on Earth is called the hydrosphere.

Enveloping the water and solid parts of Earth's surface is another relatively thin sphere comprised of gas. This is the atmosphere. All the events we call weather take place here. Another sphere, the thinnest of all, intermingles with Earth's rock, water, and air. An estimated 30 million different kinds of living things exist. This is the biosphere.

Earth's last major sphere is the transition zone between the planet and outer space. This sphere starts at the upper levels of the atmosphere, where the remaining atoms of gas are so widely spaced that no life can exist. It then stretches out thousands of miles from Earth. This sphere is the hardest to describe because its thickness can change. Furthermore, it appears empty, but it actually contains small amounts of gas. It is crisscrossed by radiation of all kinds and by intense magnetic fields. Here the ghostly light seen by *Discovery*'s crew appeared. It's here Earth first interacts with the Sun's energy. This is Earth's outer atmosphere, where planet Earth meets outer space. You are about to learn its story.

The gases in Earth's outer atmosphere are primarily hydrogen and helium at extremely low densities.

CHAPTER 1

THE EDGE OF SPACE

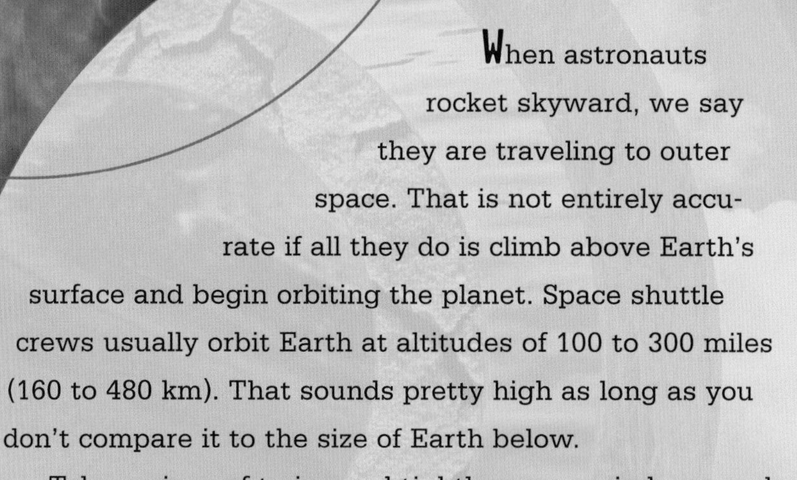

When astronauts rocket skyward, we say they are traveling to outer space. That is not entirely accurate if all they do is climb above Earth's surface and begin orbiting the planet. Space shuttle crews usually orbit Earth at altitudes of 100 to 300 miles (160 to 480 km). That sounds pretty high as long as you don't compare it to the size of Earth below.

Take a piece of twine and tightly wrap a circle around a world globe. If Earth were the size of that globe, a tiny space shuttle would orbit no higher above the globe than the outer edge of the twine! Since Earth's atmosphere reaches out toward space 10,000 miles (16,000 km) or more, space shuttles actually orbit Earth within its atmosphere.

As the Sun sets over the Sahara Desert, the thinning of Earth's atmosphere is seen in this picture taken from space by the crew of the Space Shuttle STS-101 mission.

To actually reach outer space, an astronaut has to travel beyond the atmosphere to the region between the planets. Outer space is the void between the rocky and gaseous planets and the superhot stars. Except for subatomic particles such as electrons and protons, ions (atoms that have an electric charge), and rocky or icy dust fragments, outer space is almost completely empty. That's why it's called space.

The place we usually think of as space, where astronauts orbit Earth, is really just a continuation of Earth's atmosphere. It is a vast zone surrounding the planet. It consists of extremely thin gases and magnetic field lines that pass from Earth's interior out into deep space beyond the Moon.

The material present in Earth's outer atmosphere is about as close to nothing as you can get. The amount of matter there is difficult to guess. One estimate states that if you took all the air molecules above 1,000 miles (1,600 km) and clumped them together, they would equal the number of molecules found at sea level in a single cube of air slightly less than 0.5 inch (1 cubic centimeter) on a side. Still, that's a lot of air molecules—about 20 billion billion!

INNER AND OUTER LIMITS?

Where does the outer atmosphere begin, and where does it end? There are no easy answers to these questions. The outer atmosphere may begin as low as 200 miles (320 km) or as high as 470 miles (750 km). It may end at 1,000 miles (1,600 km), 10,000 miles (16,000 km), or higher. There are many reasons why the numbers are uncertain.

One reason for not being able to come up with definite limits for the outer atmosphere has to do with the scientists who study it. There are many different kinds of science, and scientists don't always agree. A scientist's job is to try to understand the meaning behind observations and data. Depending on what a particular scientist is looking for, that scientist may define the limits of the outer atmosphere differently than another scientist would. This is something like looking at the world with sunglasses on. If you put on rose-colored lenses, everything you see will have a rosy tint to it. If you wear brown or blue lenses, the world will take on brown or blue tones. Something similar happens with scientists. The tools they use are like the sunglasses. They see and record different things.

Different kinds of scientists have devised different systems for dividing up the atmosphere, based upon their interests. The layers they identify have their own special sets of characteristics. Meteorologists (scientists who study weather) divide the atmosphere into layers by physical characteristics such as temperature and pressure and types of weather. Physicists (scientists who study forces and energy) focus on the electrical properties of the atmosphere. Physicists have discovered that the higher you go, the more electrically charged the atmosphere becomes. Atmospheric chemists look at the chemical reactions taking place. Rocket scientists are concerned with the density of the atmosphere so that they know where they can orbit their spacecraft.

For our study of the outer atmosphere, we will adopt the most commonly used system, which was created by meteorologists. Keep in mind that much overlap exists among diverse systems, but various books and websites may use different numbers. The different systems do not necessarily mean disagreement among scientists. They are just different perspectives.

In this system created by meteorologists, the lowest atmospheric layer is called the troposphere. Here most of

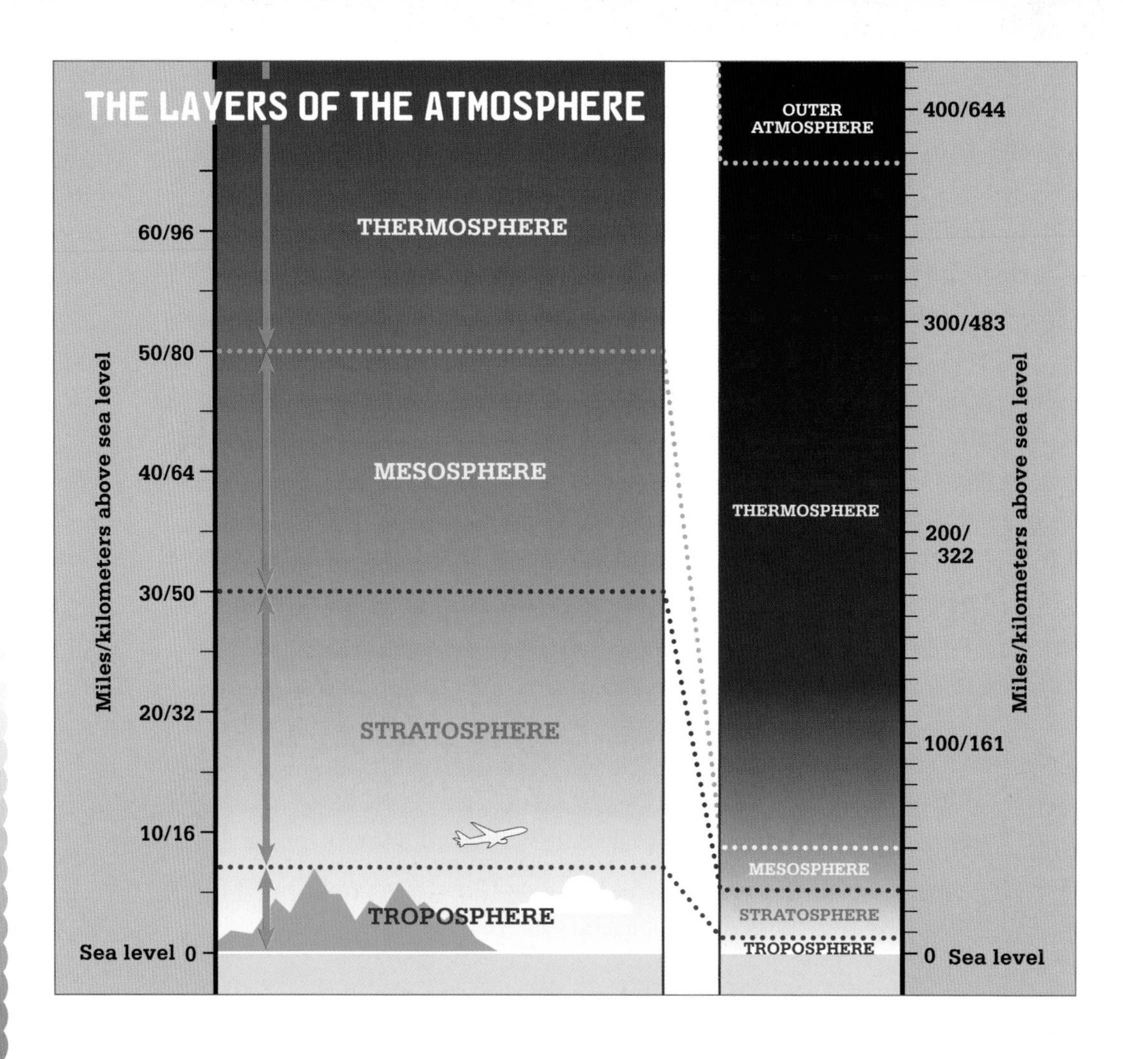

Earth's air and water vapor are found and most of Earth's weather takes place. Above the troposphere, starting at an altitude of 7 miles (11 km), is the stratosphere. The stratosphere is the calm layer of very cold air where long-distance jet airliners fly. The stratosphere stretches upward to about 30 miles (50 km).

The layer called the mesosphere is above the stratosphere. The mesosphere is where many chemical reactions take place. The upper edge of the mesosphere is about 50 miles (80 km) above Earth's surface.

THE IONOSPHERE

In your research, you are likely to come across the term *ionosphere*. This is one of those overlapping systems that physicists and other atmospheric scientists identify. The ionosphere refers to a wide region of the upper atmosphere. It stretches from within the mesosphere into outer space. The ionosphere is characterized by the presence of ions.

In the thicker reaches of the mesosphere and thermosphere and well up into the very thin outer atmosphere, individual gas atoms, such as oxygen, nitrogen, hydrogen, and helium, are exposed to intense solar radiation. This radiation creates atoms with a positive charge by knocking outer electrons out of their orbits around the nucleus. Radio waves from transmitters on Earth reflect off these scattered electrons like light reflects off a mirror. This enables radio signals to bounce around Earth, rather than simply escaping into space.

The ionosphere is not a perfect "mirror." At the wrong angle, radio waves are absorbed or pass through it. Also, bubblelike holes sometimes form in the ionosphere and break up radio waves. This can be a problem when radio stations on Earth try to communicate with a satellite or spacecraft. This can also affect the "lock" of GPS (global positioning system) navigation units. Without the lock, GPS users are unable to determine their locations on Earth.

Space weather changes have a big effect on the ionosphere. Scientists are actively studying these effects. They have divided the ionosphere into three main layers. The lowest is called the D region. Then going upward are the E and F regions. The problematic "bubbles" appear in the F region.

About 99 percent of all Earth's atmosphere lies below this edge. Once you go above the mesosphere, you are officially an astronaut and earn your astronaut wings.

In this system of layers, next is the thermosphere. Except for a few Apollo missions to the Moon, no astronauts have gone higher than the thermosphere. Up to this point, the temperature of the atmosphere has generally fallen with increases in altitude. The thermosphere gets its name from a reversal of the temperatures. Within this zone, the temperature of the few gas atoms and molecules present in the thermosphere rises as high as 3,100°F (1,700°C). Yet, if you could stick your bare hand into the thermosphere, you wouldn't notice the high temperature because the air is way too thin to feel.

Finally, there is the outer atmosphere, sometimes referred to as the exosphere. Although it stretches for thousands of miles above Earth, it contains the fewest gas atoms of any atmospheric layer. You will notice that we haven't settled on a number for how high we have to go to reach the outer atmosphere or how far out it extends. Only approximate measurements exist, for a number of reasons.

The primary reason is that the thickness of Earth's lower atmospheric layers is not constant. These air layers change in thickness all the time. They expand during the day and expand or contract with the seasons and with activity cycles of the Sun. They even change in thickness according to the positions of the Moon and the Sun in the sky. This last effect has to do with gravity.

ATOMS AND IONS

Atoms are the smallest pieces of a chemical element, such as oxygen or gold, that have all the properties of the element. Break them down further and they are no longer the same element. Atoms are made up of three types of smaller particles. Protons and neutrons are found in the center of atoms in a cluster called the nucleus. Protons have a positive charge, and neutrons have no charge. Electrons orbit around the nucleus and have a negative charge. The number of protons and the number of electrons in an atom are equal. The positive and negative charges cancel one another out. When exposed to powerful energy sources, atoms can gain or lose electrons, causing them to become ions. With enough energy, atoms can be broken apart, freeing protons and electrons.

PARTS OF AN ATOM

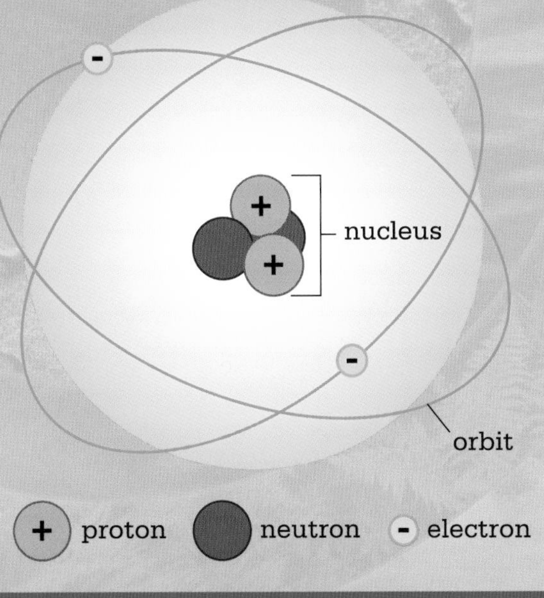

+ proton neutron - electron

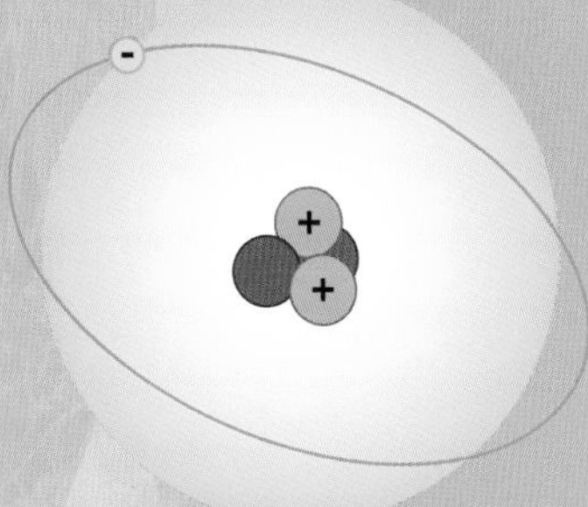

POSITIVE ION

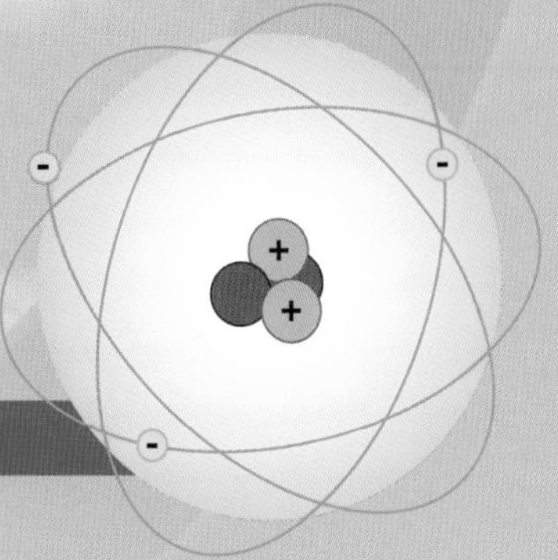

NEGATIVE ION

This is a diagram of the element helium. It has two positive protons in its nucleus that are balanced by two negative electrons orbiting its nucleus. It also has two neutrons in its nucleus.

It is similar to ocean tides, where water levels rise and fall around the world depending upon the position of the Moon and the Sun. With the expansion and contraction of the lower atmospheric layers, the outer atmosphere is pushed outward or drawn back.

The daily changes in the thickness of the lower atmospheric layers primarily occur because of heating cycles. The atmosphere is heated on the dayside of Earth. It cools off on the nightside. Temperature has a major effect on the motion of air molecules. You can see this for yourself with a simple experiment. Take an empty 2-liter (68-ounce) plastic soft-drink bottle and place it in a freezer. The bottle starts out with warm air inside, but then the freezer chills the air. In a couple minutes, take out the bottle and see what happened to it. Observe the bottle as it warms up again to room temperature.

Your observations of the bottle will help you to understand why the thickness of atmosphere layers changes between day and night. Decreasing temperature causes atoms and molecules in the bottle to move more slowly. This changes the amount of force the molecules exert when colliding with the inside walls of the bottle (slower speeds exert less force, or pressure). With less force inside due to the low temperatures of the freezer, the outside air squishes the bottle. When the inside molecules become warm again, their pressure increases and the bottle expands back to its original size.

On the dayside of Earth, increases in temperature

cause air molecules to move faster and bump against one another with greater force. The increased force causes the atmospheric layers to balloon outward against Earth's gravity. When the temperature falls during the night, the motions of the atoms and molecules slow. Less force pushes them apart, so gravity pulls the atmosphere closer to Earth's surface.

Seasonal changes also affect the air temperature. When it is summer in one hemisphere, it's winter in the other. The air over the summer hemisphere is warmer and it expands. The air over the winter hemisphere is cooler and contracts.

Still another temperature-related factor is Earth's distance from the Sun. Earth's orbit is not a perfect circle. The distance varies about 3 million miles (5 million km). Earth is

Earth's distance from the Sun varies about 3 million miles (5 million kilometers). This distance change only accounts for slight increases or decreases in the energy Earth's atmosphere receives from the Sun.

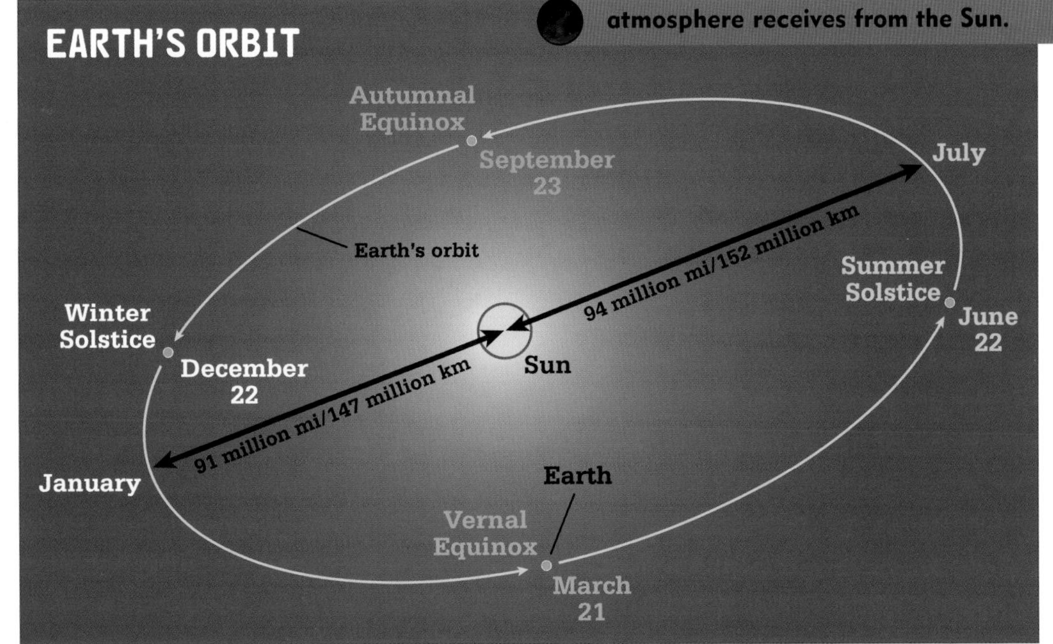

actually closer to the Sun during the Northern Hemisphere winter than it is during the summer. The change in distance slightly increases or decreases the heat energy Earth's atmosphere receives from the Sun. The temperature changes, expands, and contracts the atmosphere.

The factors that can have the greatest and most rapid effect on atmospheric layer thickness are changes within the Sun. Across tens of millions of miles (km) of outer space, solar activity (storms and explosions on the Sun) blasts out billions of tons of ions. These ions flow in million-miles (km)-per-hour streams out into the solar system. Some of those streams slam into Earth's outer atmosphere.

Solar activity affects Earth's atmosphere in a variety of ways. The big bursts of energy cause extra heating of the atmosphere, altering its thickness. They also cause the atmosphere to glow. The processes at work inside the Sun

An extreme ultraviolet imaging telescope took this image of an explosion on the Sun in 1999. The hottest areas on the image appear almost white, while the darker red areas indicate cooler temperatures. Explosions on the Sun send streams of ions into the Earth's outer atmosphere.

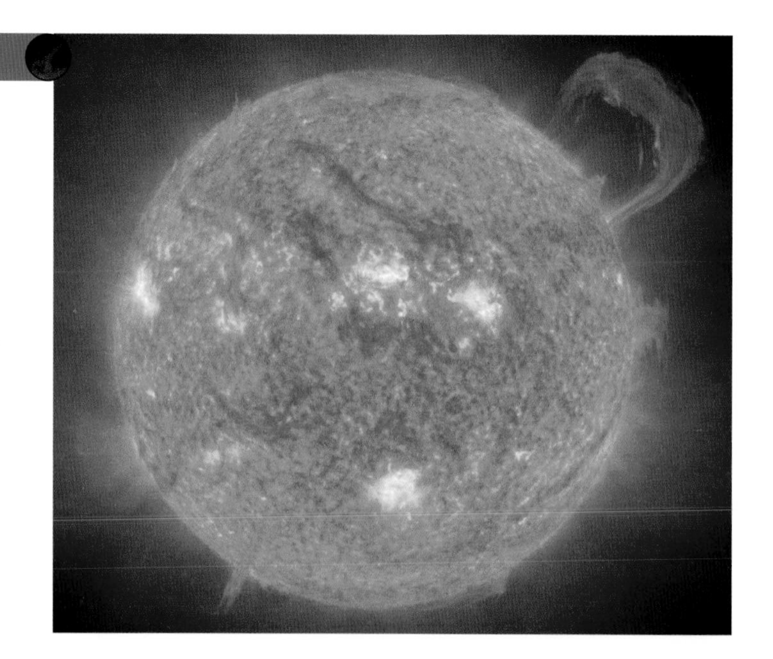

lead to the explosive ejection of particles throughout the solar system. Scientists call this space weather.

Because the thickness of the atmospheric layers changes so much, we have to pick approximate numbers for where the outer atmosphere begins and where it ends. Its lower edge (the top of the thermosphere) begins approximately 375 miles (600 km) above Earth's surface. The outer limit lies anywhere from 1,000 to 10,000 miles (1,600 to 16,000 km) above Earth. This is not a very precise measurement. It is like telling someone that your house lies somewhere between Alaska and Argentina.

The main reason for the distance range is that it is impossible to be sure just where Earth's atmosphere ends at any given time. As air thins out, it becomes harder and harder to detect. At some point, it is gone and you are in outer space. Even if you could locate the top of the atmosphere precisely, its edge will be different tomorrow. A range of 1,000 to 10,000 miles (1,600 to 16,000 km) is about as precise as it gets.

WHAT IS THE OUTER ATMOSPHERE LIKE?

Except for true outer space, the outer atmosphere is about as close to nothing as possible. It is the transition zone between air and space. It is comprised of atoms and molecules of gas that are widely spaced. These particles become more and more widely spaced farther away from Earth.

The spacing of the atoms and molecules is due to gravity. Air atoms and molecules have mass. The force of gravity acting on their mass gives them weight. The lower you go in the atmosphere, the greater the weight becomes because all the atoms and molecules are piled on top of one another. As a result, the atoms and molecules are pushed together so that very little space is between them. At sea level, the particles are so close together that they can only move about 4 millionths of an inch (0.00001 cm) before banging into one another. They collide about five billion times every second. This amounts to an air pressure of 14.7 pounds per square inch (101 kilopascals).

At the top of the pile, however, there is very little weight. One reason is simply that you are at the top. Pile up a stack of books on a table. Which is easier to lift, the book at the top of the stack or the book at the bottom of the stack?

Another reason there is very little weight is that the force of Earth's gravity decreases the higher you go. In the typical orbital range of the space shuttle, gravity diminishes to about 92 percent of the gravity at Earth's surface. At 1,000 miles (1,600 km), gravitational force diminishes to about 64 percent of surface gravity. This permits atoms and molecules to be widely spaced. Outer atmosphere atoms and molecules are so widely spaced that they may travel 1,600 miles (2,600 km) in twenty minutes before hitting another atom or molecule.

ALMOST NOTHING STILL ADDS UP

You can't see or feel them, but the atoms and molecules in the outer atmosphere are still a "force" to be reckoned with. Spacecraft and satellites orbiting Earth through the outer atmosphere continually collide at high speed with the widely spaced atoms and molecules. Each collision causes a tiny amount of drag.

You wouldn't think a 100-ton (91-metric-ton) spacecraft would be affected much, but the drag mounts up over time. Imagine that an astronaut inside the shuttle shuts off the air-moving fans so that the cabin air is very still. Then a pen is placed next to the back wall of the cabin. The pen would very slowly drift to the front wall. In several minutes, the pen would travel the short distance.

The pen would move by inertia. Outside, air collisions drag on the shuttle, but the pen inside is not affected. That makes it drift forward. Inertia is the same force that causes a soft drink to slosh around inside a cup in a moving car. Put on the brakes, and the liquid keeps moving.

In time, the drag slows a spacecraft enough that it is in danger of dropping out of orbit. A thrust from a rocket engine will speed it up again, and everything will be fine. However, if the spacecraft doesn't have a rocket engine to give it a boost, its days are numbered. You may have read about the first U.S. space station, *Skylab*. After *Skylab* was abandoned in orbit in 1974, outer atmosphere drag gradually slowed the station. It finally fell from orbit in 1979.

When molecules collide, they bounce off one another in different directions. Some bounce upward. If they are traveling fast enough when they bounce upward, they may escape the pull of Earth's gravity and travel out into the solar system. They have to be going at a good clip for that to happen—about 6.95 miles (11.2 km) per second.

The primary atoms and molecules in the outer atmosphere are the gases hydrogen, helium, and oxygen. This is very different from the troposphere, where nitrogen makes up about 80 percent of the air. (Most of the rest is oxygen, water vapor, argon, and carbon dioxide.) In the lower regions of the outer atmosphere, only a small portion of the gases are ions. These atoms have lost or gained an electron. This leaves the atoms with an electrical charge. As you go higher in the atmosphere, there are fewer and fewer atoms, but more of them are electrically charged, or ionized.

There is much more to the outer atmosphere than just thinly spaced atoms of gas. Earth's magnetic field extends from Earth's core through the outer atmosphere into deep space. In addition, many things pass through the outer atmosphere on their way toward Earth. These things include meteors—tiny, solid fragments of dust, rock, and metal—and different kinds of space radiation. Meteors, if they travel on a steep path toward Earth, heat up quickly. This is due to friction with air molecules that increase in density the lower the meteors fall. For a brief moment,

Meteors streak past stars in the night sky near Amman, Jordan, during the Perseid meteor shower. The Perseids happen every August, when Earth passes through a stream of space debris left by Comet Swift-Tuttle.

these meteors emit brilliant streaks of light that we call shooting stars.

Space radiation, also referred to as space weather, consists of ions, subatomic particles, and electromagnetic waves blasted into space by explosions from the Sun and other stars. Much of this radiation is very energetic and can penetrate the walls of spacecraft. Astronauts can become very sick if exposed to enough of it. Like forecasters on Earth that tell you what the day's weather will be like, a special group of forecasters monitor and predict changing levels of space weather. You might think only astronauts need to worry about space weather, but you would be wrong. Space weather affects all of us.

TAKING CHANCES

In addition to riding on the top of a highly explosive rocket and traveling to the deadly environment of space, Apollo astronauts of the 1960s and 1970s took their chances with space weather. Space weather is the charged particle radiation present at any moment. Like weather on Earth, space weather can range from mild to deadly.

No major solar storms erupted while astronauts were traveling to and from the Moon. Had one erupted, the crew could have become very sick from radiation. They would have been very weak and subject to vomiting and long-term organ damage. Such radiation could severely damage blood-forming marrow in their bones, requiring them to receive blood transfusions and marrow transplants. Their eyes also would have suffered damage that would lead to cataracts fogging their lenses and reducing their vision.

Future astronaut crews will spend a longer time on the Moon than the Apollo crews did decades ago. Trips to Mars will be much longer than those to the Moon. Crews will be more likely to be exposed to dangerous space weather. Scientists are working hard to find ways of protecting future space crews from intense space weather. Several strategies are being studied. The simplest is to limit the time astronauts are in space. Faster rockets decrease travel time and the total exposure to radiation. Scientists are studying the idea of nuclear-powered rockets. This could shorten a trip to Mars by several months.

Another strategy is to add more radiation shielding to the spacecraft. Scientists have to be careful, though. If they add too much extra weight, they will have to build a more powerful rocket. However, materials rich in the element hydrogen are especially good for shielding radiation.

The *Apollo 16* rocket launches on April 16, 1972, from the Kennedy Space Center in Cape Canaveral, Florida. Apollo landed on the Moon on April 20 and returned to Earth on April 27. The astronauts on Apollo didn't encounter any major solar storms on their mission.

National Aeronautics and Space Adminstration (NASA) scientists are examining a kind of plastic, similar to grocery store bag material, for spacecraft construction. Thick layers of it might replace heavier metals for the walls of spacecraft. The material would save weight and create a better radiation shield. Other potential strategies include new space foods high in antioxidants (chemicals that help to repair radiation damage) and new medicines that minimize radiation's effects.

CHAPTER 2

THE SUN HAS ITS MOMENTS

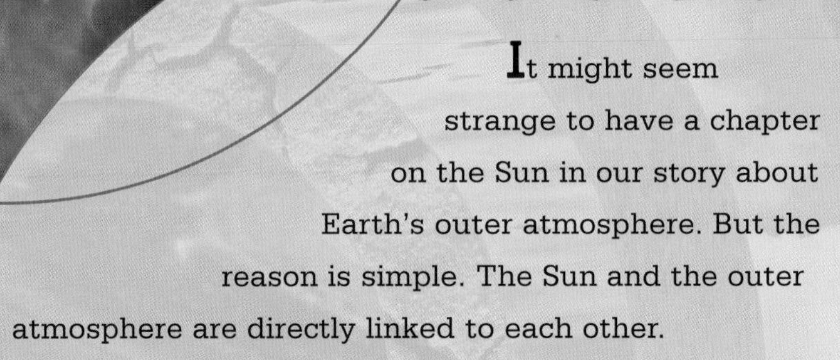

It might seem strange to have a chapter on the Sun in our story about Earth's outer atmosphere. But the reason is simple. The Sun and the outer atmosphere are directly linked to each other.

The amount of energy produced and sent out by the Sun into the solar system changes constantly. When the energy arrives at Earth, it encounters the outer atmosphere first. Earth's outer atmosphere and the magnetic field lines that cross it and extend into space are Earth's front line. They are the part of Earth that interacts first with the light energy and the charged particle radiation emitted by the Sun into the solar system. This is what space weather is all about.

We know the Sun to be the center of our solar

system. Its gravity holds Earth and all the other bodies in the solar system in orbit around it. The Sun is the source of incredible energy that floods outward. This energy warms Earth, drives its atmosphere engine, and keeps the oceans from freezing. The energy in sunlight is captured by plants and used in the photosynthesis process to make food for growth. The energy is passed on to the animals that eat the plants and to the animals that eat the animals that ate the plants.

In spite of all the wonderful properties of the Sun, it does not have entirely positive effects on Earth. The Sun ejects streams of deadly radiation that travel billions of miles into space. Earth's outer atmosphere and magnetic field really come in handy when those streams arrive at Earth. They take the hits first. To understand these effects on the outer regions of Earth, we need to know more about our Sun and what it produces.

OUR STAR

The Sun is a gigantic ball of gas approximately 864,000 miles (1,390,000 km) in diameter. It is made up of about 76 percent hydrogen and 22 percent helium. The remaining few percent are other elements such as iron and carbon. All told, more than seventy different elements have been identified in the solar atmosphere.

Hydrogen is the simplest of all the atoms that make up our world. It has a nucleus containing a single proton.

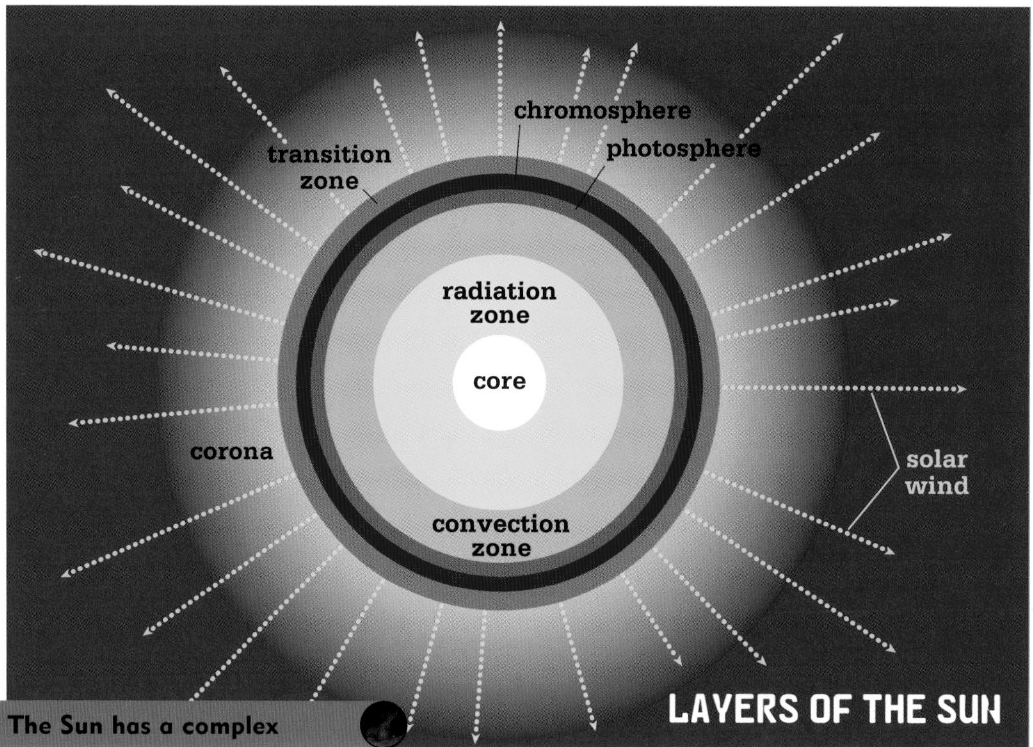

The Sun has a complex structure that accounts for the creation of massive amounts of energy that floods our solar system.

Orbiting around it is a single electron. Though only about two-thousandths the size of the proton, the electron carries an equal but opposite negative charge. The two charges balance each other so that the charge of the hydrogen atom is neutral. Helium is a more complex atom. It has two protons and two neutrons in its nucleus. Orbiting around the nucleus are two electrons.

Deep within the heart of the Sun, nature's most powerful process is hard at work manufacturing energy for the solar system. The process is called fusion, and it takes place under tremendous heat and pressure. This heat and pressure is provided by the sheer mass of the Sun. The Sun weighs about 333,000 times as much as the entire planet Earth.

Approximately 400,000 miles (640,000 km) beneath its blistering surface, the weight of all the gas piled on the Sun's core exerts a pressure of about 3,500 billion pounds per square inch (24 trillion kPa). That's about 250 billion times Earth's air pressure at sea level. The temperature is amazing too. At the core, the temperature climbs to about 27,000,000°F (15,000,000°C).

The core's heat and pressure fuse hydrogen atoms together to make atoms of helium. Fusion requires four atoms of hydrogen to make one atom of helium. The atoms have to go through three stages to complete the process. In the end, every 2.2 pounds (1 kg) of hydrogen entering the fusion process becomes 2.185 pounds (0.9911 kg) of helium. Approximately 0.015 pounds (0.0068 kg) is turned into energy that is ultimately released by the Sun into the solar system.

The fusion reaction within the Sun combines hydrogen to create helium while releasing huge amounts of energy that floods the solar system.

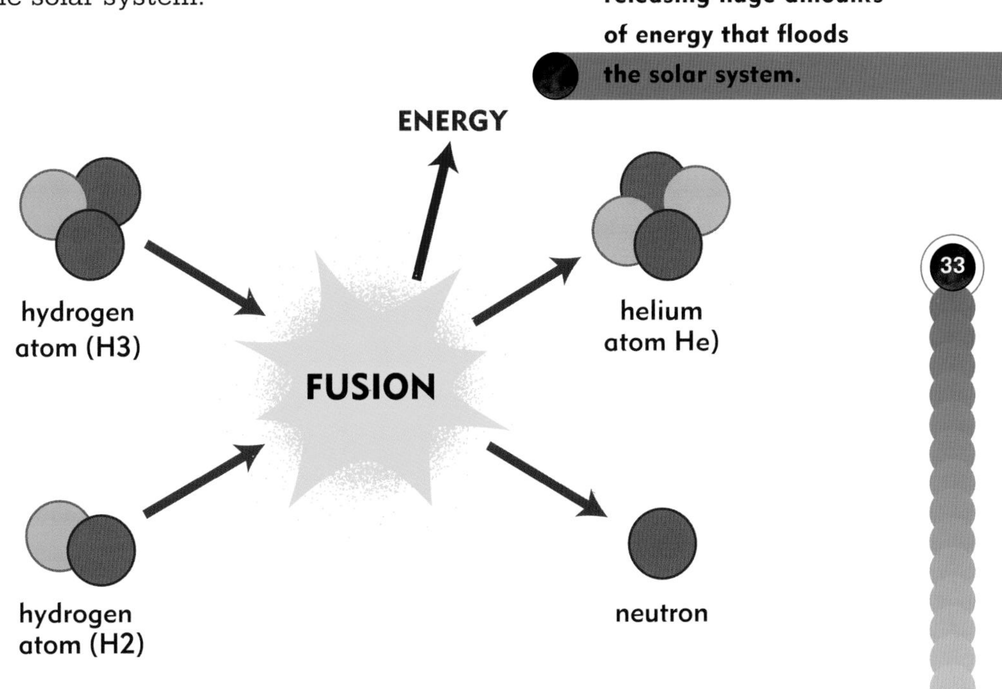

Of course, the Sun doesn't fuse just a few pounds of hydrogen at a time. Each second, it fuses about 700 million tons (630 million metric tons) of hydrogen into about 695 million tons (630 million metric tons) of helium and huge amounts of energy. The amount of hydrogen being converted every second is staggering. Fortunately, the Sun is so large that its hydrogen supply will last several billion years before it runs out.

Most of the energy produced in the Sun's fusion process is a dangerous and invisible kind of radiation called gamma rays. Gamma rays are fatal if you are exposed to enough of them. The inside of the Sun is absolutely black because gamma rays cast no light. Fortunately, the gamma rays have a long and tortuous path before they reach the Sun's surface. The trip can take up to a million years. Meanwhile, most of the gamma rays interact with electrons and protons within the Sun to produce less dangerous radiation, including photons (particles) of visible, infrared, and ultraviolet light.

The Sun's surface is an inferno. Much cooler than the core, it still approaches a searing 10,000°F (5,500°C). The surface seethes with bubbling, planet-sized storms of hot, glowing gases. Occasional eruptions take place that toss out massive clouds of hot gas. These clouds are warped into large hooks and broad arches. The storms are controlled by internal magnetic forces produced by the movements of hot gases in the interior. The forces extend from the Sun's surface in a huge magnetic field far out into space. These forces hook and loop the gases erupting from the surface.

From time to time, magnetic forces slow the outward movement of energy on some areas of the surface. The slowing causes irregular dark patches, called sunspots, to appear on the surface. The patches are actually very bright, but their slightly lower temperatures cause them to appear dimmer. (Unless a sunspot is very large, it can be seen only with a specially equipped telescope. Never look directly at the Sun, especially with a telescope or other optical instrument. Your eyes can become severely damaged.)

Solar astronomers have been tracking the number of sunspots for hundreds of years. They discovered that the numbers increase and decrease dramatically on roughly an eleven-year cycle. During active sunspot periods, the Sun experiences higher numbers of solar storms. Pent up energy from the sunspots explodes

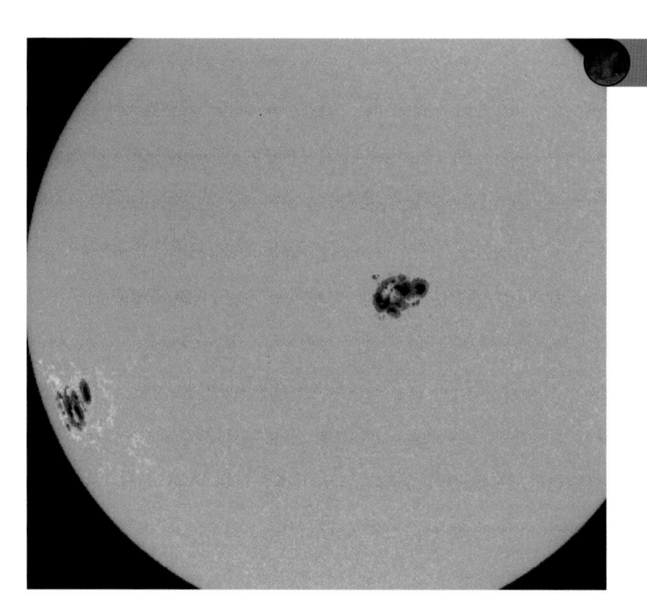

This photo shows sunspot clusters on the surface of the Sun. These sunspots produced the largest solar storms ever recorded. Solar astronomers carefully track the numbers and positions of sunspots to predict solar storms and changes in space weather.

outward as big flares of superhot gas. The flares release vast numbers of charged particles that race outward at speeds of over 1 million miles (1.6 million km) per hour. The particles are mostly protons and electrons that were ripped from the atoms of solar gases.

Solar storms aren't the only source of these particles. Even during quiet periods, the Sun's atmosphere, or corona, sheds its charged particles in a continuous flow. The temperature and consequently the speed of the particles is so high that the Sun's gravity is not strong enough to hold them. The particle flow, again mostly protons and electrons, spreads out through the solar system to engulf the planets. The flow is called the solar wind.

Solar wind is not unlike the wind in Earth's atmosphere, which is made up of moving atoms of gas. It is not a steady flow. It can range in speed from 1 million to 2 million miles (1.6 to 3.2 million km) per hour from different parts of the corona. Like a turbulent river on Earth that has water of different speeds flowing into it from tributaries, the solar winds from different areas interact with one another.

Solar winds take days to arrive on Earth, which orbits the Sun 93 million miles (150 million km) away. Sunlight streams across this distance at a speed of 186,000 miles (300,000 km) per second. It takes light about eight and a half minutes to make the trip.

Sunlight consists mostly of infrared and visible light.

LOVELY DAY FOR A SAIL

Long ago, science-fiction writers took note of the pressure created by the photons of light as they flow outward from the Sun. The pressure is minimal, especially when compared to the pressure exerted by Earth's wind. But writers proposed using it as a free source of space propulsion.

In these stories, future astronauts erect flimsy sails several miles (kilometers) wide to capture the Sun's photons. Their small spacecraft are attached to the sails by lines and dragged along. By rolling up a part of the sail to make it smaller or by unfurling it to make it as large as possible, solar sailors controlled the speed of their spacecraft. Trips to distant planets took a long time, but when their solar spacecraft got there, they still had most of their rocket fuel left for the return trip. Sometimes, rocket scientists consider using solar sails. One day, they may actually build solar sailing ships.

Infrared light transports is the radiation from the Sun that transports the Sun's heat to Earth. When you warm yourself by a campfire, it is infrared radiation that you feel. Visible light consists of the colors of the rainbow. Together, infrared and visible light make up about 90 percent of all the radiation coming from the Sun. The remainder is radio waves and dangerous forms of radiation such as ultraviolet light (sometimes called black light), X-rays, and gamma rays.

THE ELECTROMAGNETIC SPECTRUM

The light given off by the Sun and stars is a blending of many different kinds of radiation. The entire range of radiation is called the electromagnetic (EM) spectrum. It gets its name because radiation is associated with electric and magnetic fields.

The most familiar part of the electromagnetic spectrum is the visible bands of red, orange, yellow, green, blue, indigo, and violet light. These are the rainbow colors. In addition to visible light are radio waves, infrared light, ultraviolet light, X-rays, and gamma rays. Visible light is in the middle of this range, between infrared and ultraviolet.

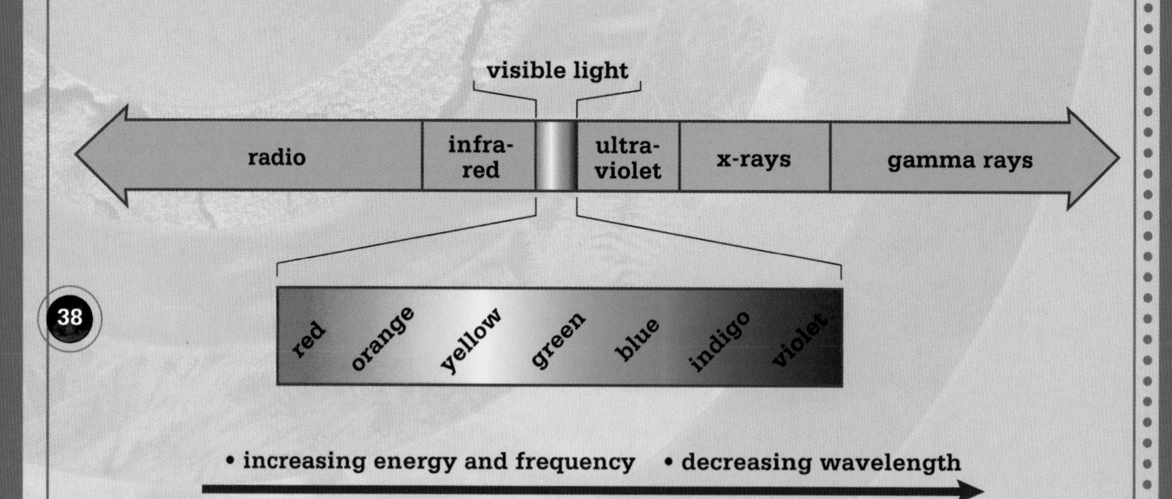

Scientists arrange the different forms of radiation of the EM spectrum in a chart. Normally, the chart starts with radio waves on the left and ends with gamma rays on the right. Waves of radiation are similar to water waves in that they have wavelengths. A wavelength is the distance from one wave crest to the next.

Radio waves have the longest wavelengths, ranging from thousands of miles to just a few inches wide. Infrared wavelengths are smaller, and visible light waves are smaller still. It would take fifty visible light waves, arranged end to end, to equal the thickness of a sheet of household plastic wrap. The wavelengths of gamma rays are smaller than the diameter of an atom.

The energy carried by waves changes with their wavelengths. The longest radio waves carry the least amount of energy, and the short gamma rays carry the most. You don't want to be hit with lots of gamma rays, because they can kill living cells. Exposure to an intense dose of gamma rays is fatal.

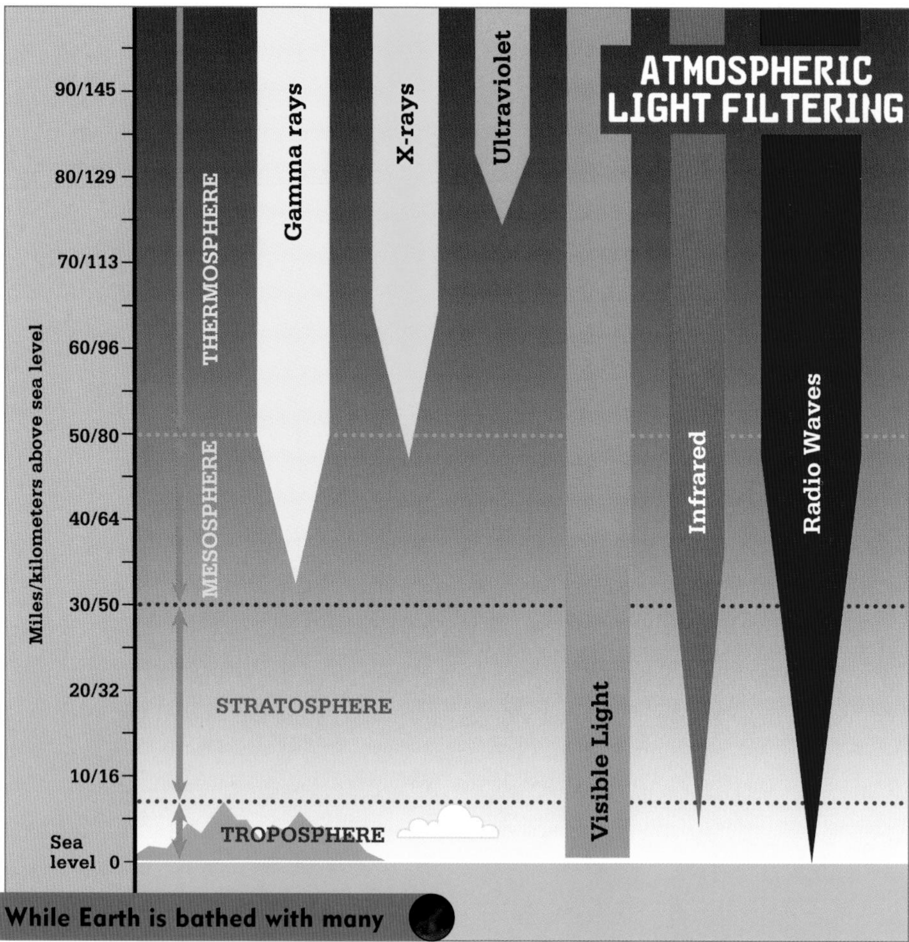

While Earth is bathed with many kinds of radiation from the Sun and from other stars, most of the dangerous radiation is filtered out by our atmosphere.

When the Sun's light and the solar winds impact Earth's outer atmosphere and its magnetic field, most of the ions either get captured by Earth's magnetic field or get deflected away. The atmosphere below it absorbs ultraviolet light, X-rays, and gamma rays. When radiation strikes matter (such as the gas in Earth's atmosphere), some or all of its energy may be transferred to the matter as heat. The collisions dissipate the radiation. Visible light and some infrared, ultraviolet, and

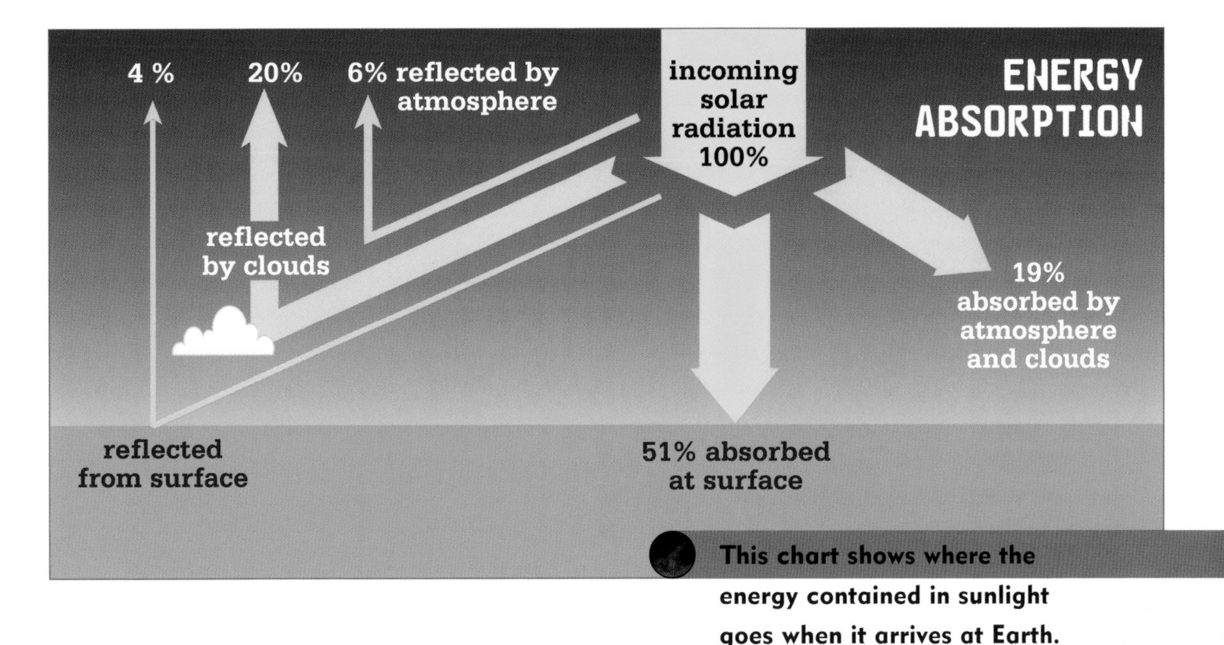

This chart shows where the energy contained in sunlight goes when it arrives at Earth.

radio radiation make it all the way to the surface. These are the safest portions of the Sun's radiation. Although, too much exposure to ultraviolet rays from sunlight (such as excessive suntanning) can lead to skin cancer.

THE HALLOWEEN STORMS

The Sun has some dramatic moments. For two weeks around Halloween in 2003, the Sun unleashed the most powerful solar storms ever recorded. An estimated seventeen major flares burst out into the solar system. Space weather scientists predicted the effects of the storms would push out the Sun's heliopause an additional 400 million miles (640 million km).

The heliopause is the outer edge of the Sun's domain, where the solar wind meets interstellar space. This is the farthest away from the Sun that its ions can be felt. The

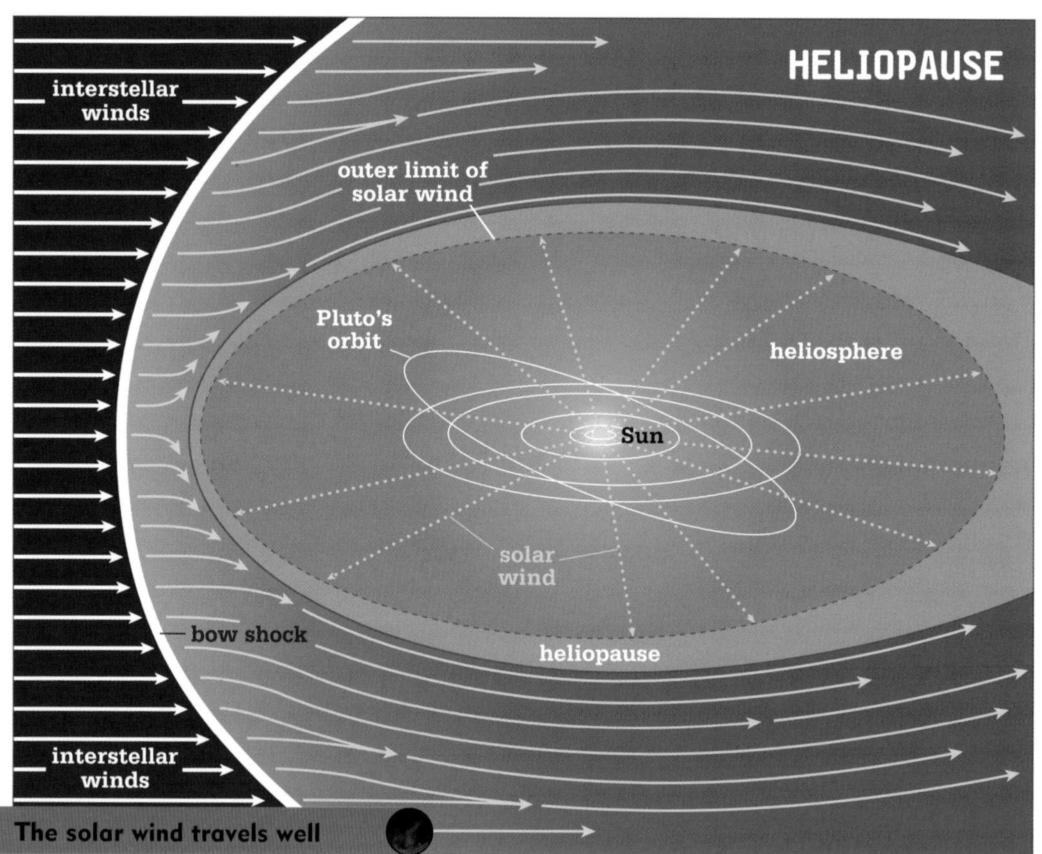

The solar wind travels well beyond the orbit of Pluto and interacts with winds from other stars in our galaxy.

heliopause is estimated to be two to four times as far away from the Sun as Pluto's average orbital distance. This means it is anywhere between 8 and 15 billion miles (13 to 24 billion km) from the Sun.

When the Halloween storms began erupting, an armada of spacecraft started collecting data. The *Ulysses, Cassini, Mars Odyssey,* and *Voyager 1* and *2* spacecraft, plus a string of solar and Earth satellites, began watching for the effects of the storms. Except for astronauts on the International Space Station in low-Earth orbit, who remained inside to stay safe, no humans were at risk.

However, human technology was at risk. Massive streams of ions erupting from the Sun can knock out electronic systems on hundred-million-dollar satellites in orbit. Radio and cell phone communication can be interrupted. On occasion, electric power grids can also be affected.

By the time the first of the storm's particles reached Earth's outer atmosphere, emergency procedures were enacted. Many satellite systems were temporarily switched into "safe" modes for protection. This is similar to unplugging your computer during electrical storms to prevent power surges from damaging it. Because of the advance warning, most potential damage from the storm was averted. A power grid in Sweden did take a hit. An electric blackout occurred that lasted only about an hour before the grid was back on line. It was a minor disturbance compared to what might have occurred if there had been no warning.

Vibrant color illuminates the night sky over Norway, after a massive solar flare from the Sun caused aurora borealis (northern lights) to appear all over the Northern Hemisphere.

The Sun contains about 98.8 percent of the total mass of the solar system. The planets, moons, asteroids, and comets make up only 0.2 percent of the solar system's mass.

Though a great distance away from Earth, the Sun nevertheless bathes Earth with many kinds of radiation that influence much of what takes place here. But the Sun is not alone in its influence. Radiation from other objects deep in space has its effect too. Yet another influence comes from within Earth itself—its magnetic field.

MARIE IS NO MORE

Thanks to the advance warning of the 2003 Halloween storms by spacecraft and satellites, little damage to Earth-orbiting spacecraft and Earth's power and communication systems occurred. However, a significant incident took place on the planet Mars. The U.S. *Mars Odyssey* spacecraft orbiting the planet took a big hit.

One of the instruments on the spacecraft was designed to measure the radiation environment of Mars. Mission planners would then know what to expect when the first humans arrive there. The instrument was called MARIE, or the Martian Radiation Environment Experiment. The radiation from the Halloween storms overheated and fried a power converter for the instrument. It became useless.

SOLAR BUBBLES

The Sun has a broad but faint atmosphere. For thousands of years, the corona could be seen only during total eclipses—when the Sun's bright disk is completely blocked by the Moon.

In the early 1970s, satellites designed to study the Sun were launched. They carried instruments that could simulate total eclipses. One of their discoveries was that coronal mass ejections, or CMEs, are like giant bubbles of gas that erupt from the Sun over several hours. CMEs are triggered by rapid changes in the Sun's magnetic field. The bubbles are shaped by the magnetic field lines threading through them.

In a CME, up to 10 billion tons (9 billion metric tons) of electrically charged gas bulge outward from the Sun. The gas travels across space at 4 million miles (6.4 million km) per hour. Like the solar wind, the charged gases in a CME can slam into Earth's outer atmosphere and disrupt communications and power systems on Earth. Space weather forecasters watch out for CMEs. When one occurs, the Space Weather Center sends out alerts to satellite, power plant, and communication systems operators worldwide.

On October 28, 2003, a huge solar flare resulted in a powerful coronal mass ejection of high-energy protons toward Earth. This image, collected by the Solar and Heliospheric Observatory, blots out the bright disk of the Sun so that the fainter atmosphere can be seen. The white circle shows the position of the Sun in the picture.

CHAPTER 3

DEEP DOWN BELOW

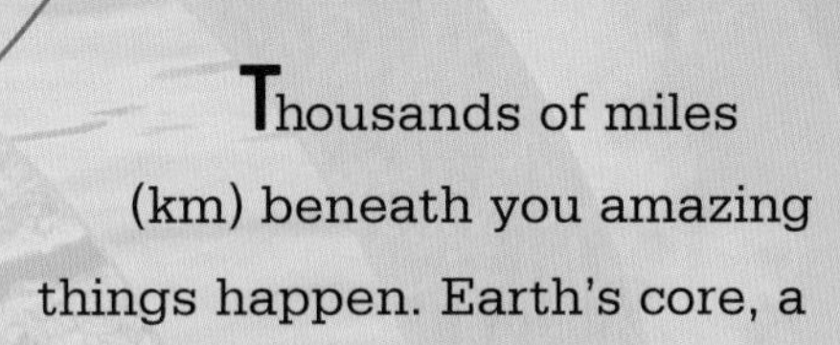

Thousands of miles (km) beneath you amazing things happen. Earth's core, a two-layer metal sphere, occupies Earth's center. The innermost part, or inner core, begins at the center and extends out about 745 miles (1,200 km). Stretching beyond the inner core approximately 1,400 miles (2,200 km) is the outer core. Together, the inner and outer cores make up only about 16 percent of Earth's total volume, but they comprise about 32 percent of the planet's total mass. That means that the core is very dense compared to the rest of the materials that make up Earth. Its density comes primarily from two elements. About 90 percent of the core is the element iron. The other 10 percent is the element nickel.

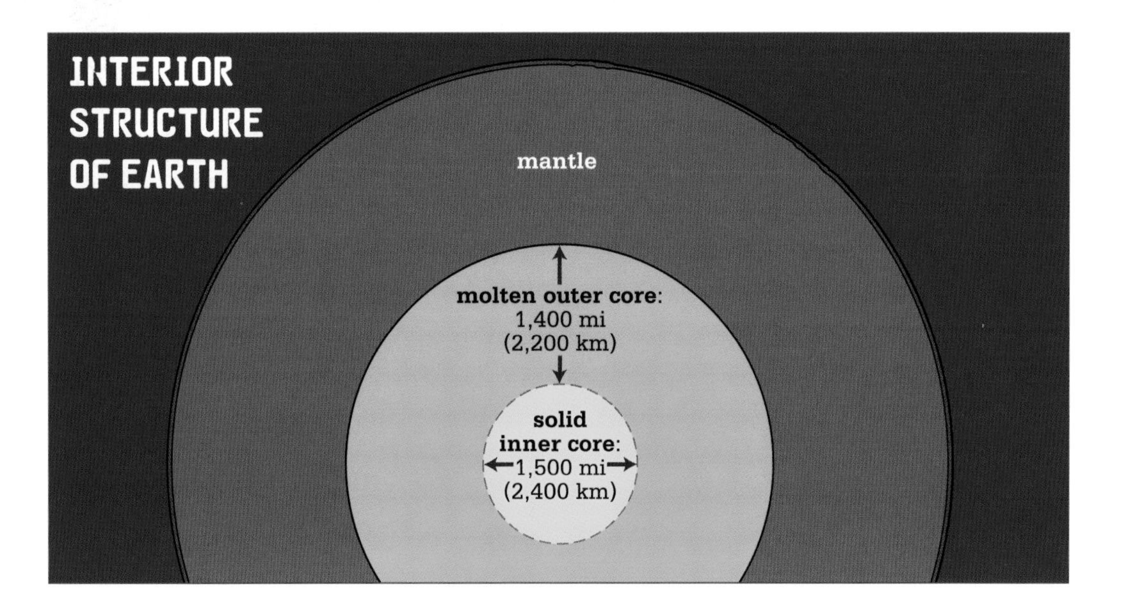

The division of the core into two layers is based on their states of matter. The inner core is solid, and the surrounding outer core is molten. The nature of the core as solid and liquid was discovered when scientists carefully measured earthquake waves as they passed through Earth.

Earthquakes create various kinds of waves. One type passes through both liquids and solids, while another passes only through solids. Liquids, such as molten iron and nickel, do not transmit the second kind of wave. Scientists were able to create a model of Earth's interior based on how the waves moved through the planet. There was a large shadow zone that the second wave type did not pass through. The shadow zone corresponded to the presence of a molten outer core.

Scientists wondered why the outer core was molten but the inner core was not. The answer turned out to be heat and pressure. The deeper you go into Earth, the greater the pressure becomes. The pressure is caused by the weight of the overlying material. Along with the increase in pressure comes an increase in temperature.

If you could descend into Earth, you would quickly notice the rising temperature. By the time you reached the boundary of the inner core, the temperature would be great enough for iron and nickel to melt. When you finally arrived at the center, the temperature would peak around 9,030°F (5,000°C). This is almost as hot as the surface of the Sun.

Although the temperature of the inner core is more than hot enough to melt the iron and nickel, these metals are solid due to the extreme pressure. When most materials melt, they expand in volume. The pressure on the iron and nickel in the inner core prevents them from expanding from the solid to the molten state.

The extreme temperature in Earth's inner core is caused by a few different factors. One cause is the heat generated when Earth and the solar system were created by the collapse of a giant cloud of gas. In its early days, Earth was entirely molten. Billions of years later, it is still cooling off.

Another part of Earth's internal heat comes from the decay of radioactive (unstable) forms of some elements. These elements gradually shed protons or neutrons from their nuclei, causing them to change into other elements. Radioactive uranium gradually transforms itself into lead. Radioactive carbon becomes nitrogen. Depending upon the element, the change can take thousands, millions, or billions of years. One of the by-products of this radioactive decay is heat.

A third source of heat is the daily gravitational tug of the Moon. The Moon and Earth pull on each other, which results in ocean tides. The pull also causes very slight tides within Earth. The friction caused by shifting rocks contributes to Earth's internal heat.

In the previous chapter, we examined the Sun's powerful influence on the outer atmosphere. As we will see, Earth's outer core, thousands of miles beneath the outer atmosphere, also has a powerful influence. The influence has to do with electricity and magnetism.

ELECTROMAGNETISM

You have probably investigated electricity and magnets in science class. Some of your experiments may have involved building electric circuits to light small bulbs or sprinkling tiny bits of iron

A simple electromagnet is made by wrapping wire around a nail. Current, running through the wire, causes the nail to become magnetized.

ELECTROMAGNET

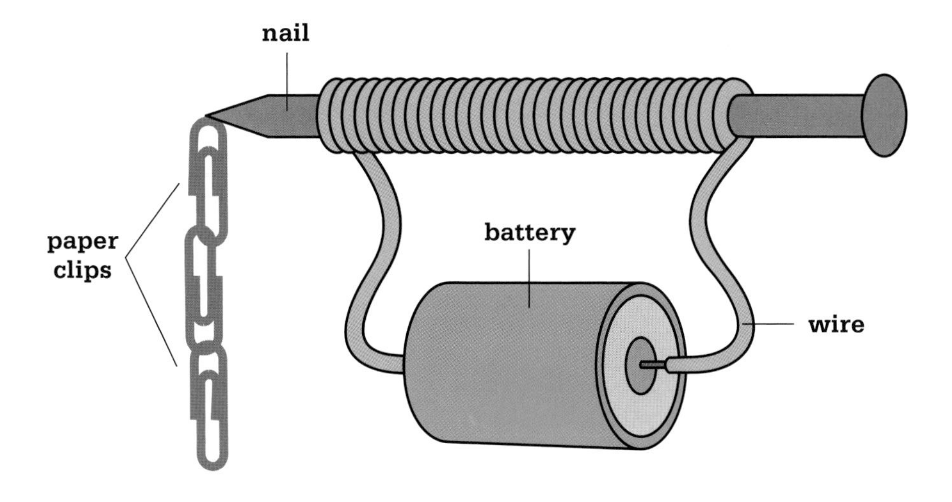

around a bar-shaped magnet to see its magnetic field.

One of the more important science class experiments involves wrapping a coil of wire around an iron nail. Then a battery is connected to the two ends of the wire. The electric current running through the wire magnetizes the nail, enabling it to pick up paper clips. When the current stops, the magnetism goes away and the paper clips fall. The experiment is important because it demonstrates that electricity and magnetism are related. Electric currents create magnetic fields and vice versa.

In another important experiment, a magnet is moved through a coil of wire and it generates an electric current. The experiment shows that magnetism creates electricity and electricity creates magnetism. This is called electromagnetism. It is used to run generators, electric motors, lightbulbs, and everyday electric appliances. Electromagnetism is also at work within Earth. Its effects on the outer atmosphere and beyond are powerful.

Earth's magnetic field reaches up from the core through all the overlying surface rock. It extends hundreds of thousands of miles out into space. It is the magnetic field of Earth that causes a magnetic compass needle to point in the general direction of north so that we can find our way. It is the magnetic field that protects Earth's surface from the constant flow of dangerous radiation from the Sun.

Where does Earth's magnetic field come from? It was first thought that iron in Earth's interior was permanently magnetized, like a bar magnet. That idea was rejected

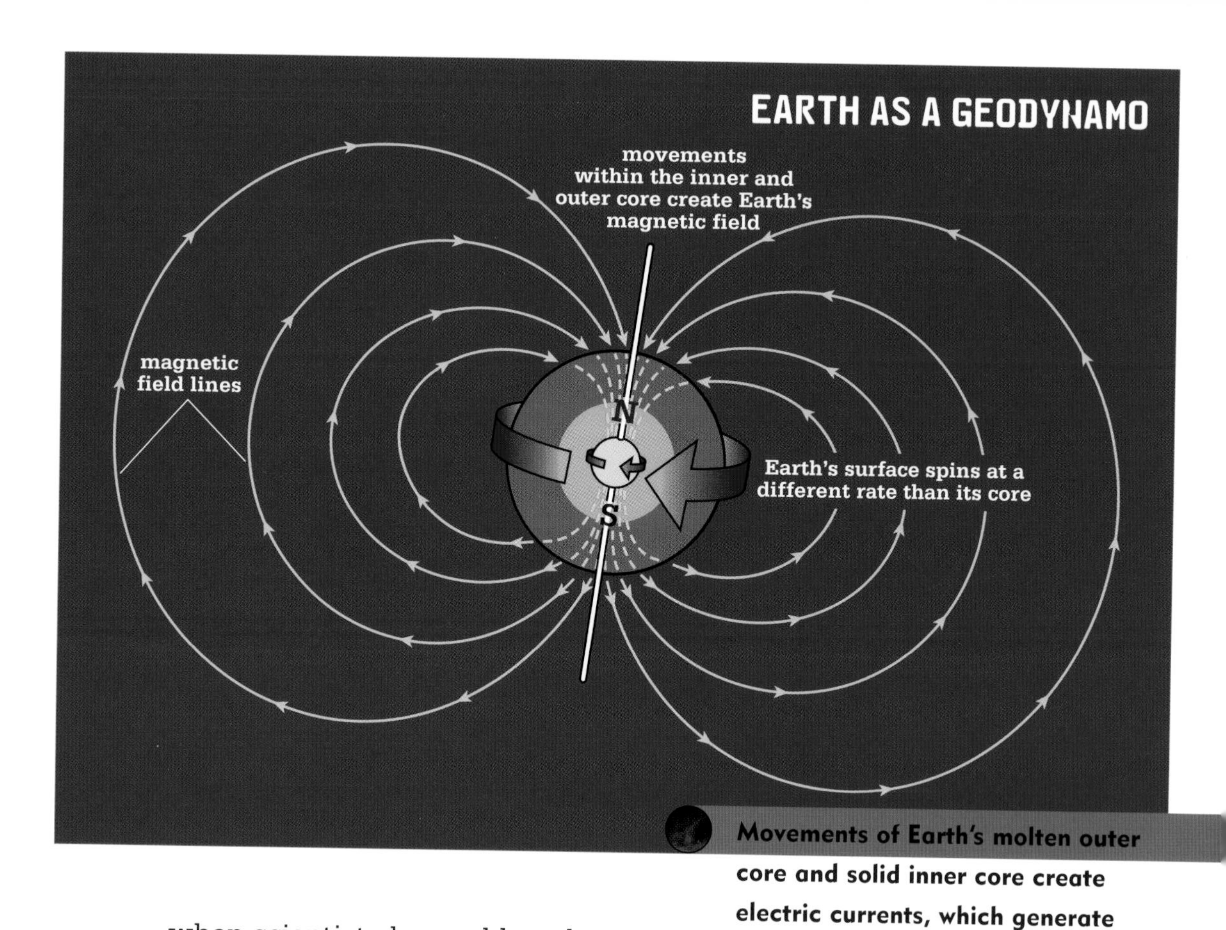

Movements of Earth's molten outer core and solid inner core create electric currents, which generate Earth's magnetic field.

when scientists learned how hot Earth's interior is. If you heat a permanent iron magnet beyond a certain temperature called its Curie point, the iron loses its magnetism. Earth's interior is far too hot to be a permanent magnet. So how does it generate its magnetic field? The field comes from electric currents in the molten inner core.

The molten iron of the inner core is in constant motion. Convection currents (such as the currents in a boiling pot of soup) cause the molten iron to rise and fall in big swirling arcs. The superhot material deep in the inner core expands and slowly floats up to the top of the inner core. There, heat is transferred to the overlying mantle. This causes the

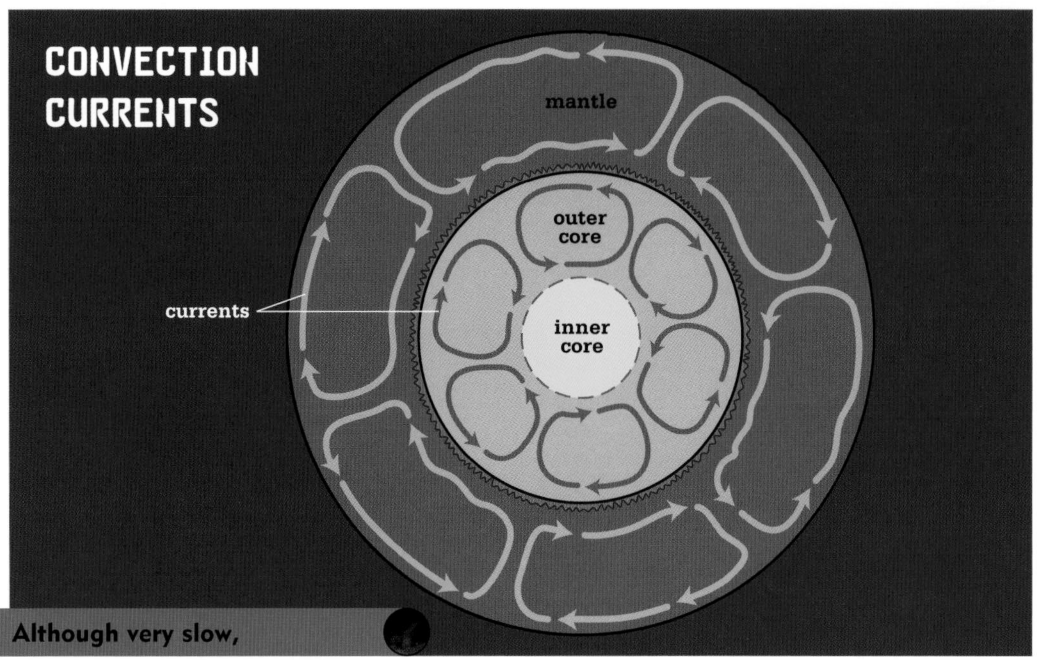

Although very slow, movements of material within Earth's layers produce Earth's sheltering magnetic field.

molten material to cool and contract. Then it falls back down to warmer regions to be heated again.

While the molten material is rising and falling, Earth is rotating. The effects of rotation and rising and falling convection currents cause a flow of molten material in the outer core. The great heat causes atoms to become ionized, and the flow of these charged particles through the inner core creates powerful electric currents. These currents, in turn, create the magnetic field that extends out into space.

All this activity is referred to as the dynamo effect. Dynamo is the name that was given to the first mechanical electric generators. The dynamo effect of Earth is self-sustaining. The flow of molten material through Earth's magnetic field generates an electric current that, in turn, creates the magnetic field.

THE DYNAMIC DYNAMO

Scientists love studying Earth's magnetosphere because so many mysteries surround it. For example, Earth's magnetic field generates electric currents in the outer core, but these currents generate the magnetic field. Did the magnetic field come first, or did the electric currents? As with all Earth systems, there is more going on.

The inner-core electric current isn't the only current. The solar wind consists of electrically charged particles. The particle flow, called plasma, is actually an electric current of particles. When the plasma reaches the vicinity of Earth's magnetic field, it interacts with the field lines and adds energy to them. Like an electric circuit in school science experiments, the plasma current forms a loop as it travels along one or more field lines into Earth. Then it returns to space along other field lines. The plasma current strengthens the magnetic field, which strengthens the outer core electric currents, which strengthen the magnetic field, and so on.

THE GEOMAGNETIC FIELD

Remember the experiment with the bar magnet and iron filings? If you have a bar magnet and some filings (or tiny fragments from some steel wool), do the experiment again and study the shape of the magnetic field. Lay a piece of paper over the magnet, and sprinkle the filings on it. The particles will be attracted to the ends of the magnet, where the magnetic field is strongest. Field lines

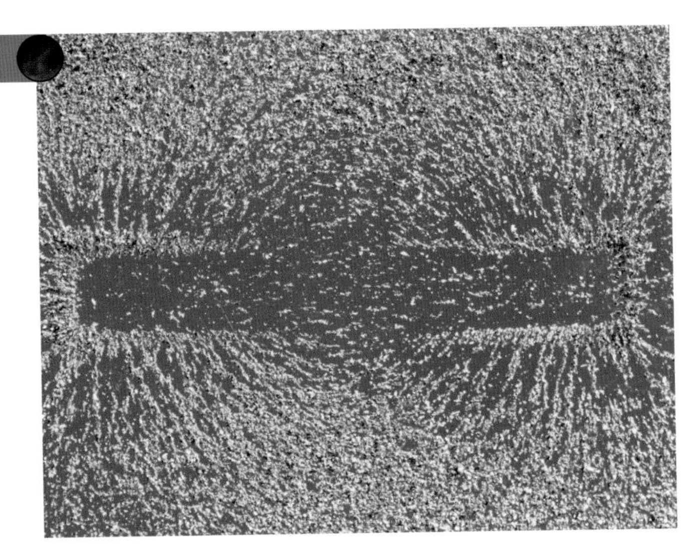

Iron filings around a bar-shaped magnet show the magnetic field of the magnet in this experiment.

become visible as loops that extend from the magnet's north pole to its south pole.

Although you can only see where the field lines intersect with the surface of the paper, they actually surround the magnet in three dimensions. The lines going above and below the magnet are not strong enough to suspend the metal filings in the air.

The reason the ends of the bar magnet are able to attract and pick up iron objects is that the field lines are closest together on the ends. This concentration of field lines produces a stronger magnetic force than in other places around the bar magnet, where the lines are farther apart. A few field lines will be open and will not loop from one end of the magnet to the other. These lines arc outward into space.

Earth's magnetic field, also called the geomagnetic field (*geo* means "Earth"), is similar to the field around a bar magnet, but there are important differences. For instance, the bar magnet's north and south poles are always on the ends of the magnet. Earth's magnetic north and south poles are not fixed. They slowly wander across Earth's surface because of the movements of the molten iron in the inner core. Rarely are the magnetic poles close to the North and South Poles that are determined by Earth's rotational axis.

Furthermore, the magnetic north and south poles are not directly on opposite sides of Earth. The magnetic north pole is about 10 degrees from the true North Pole. The magnetic south pole is about 25 degrees from the true South Pole. This is because the center of Earth's magnetic field is a bit out of alignment with the center of the planet. It is off by about 300 miles (500 km) from Earth's geographic center.

The misalignment produces a lowering of the magnetic field strength over the South

Earth's north and south magnet poles wander slowly and are offset by several degrees from the geographic north and south poles.

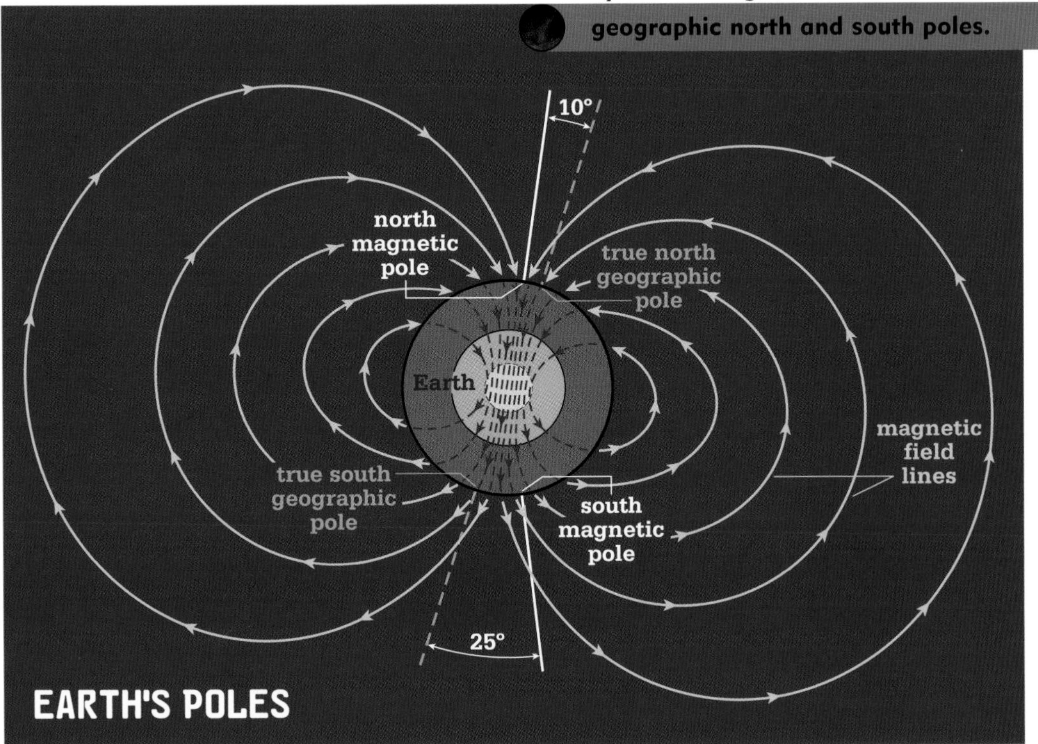

Atlantic Ocean. Here, radiation ions from space are able to penetrate the outer atmosphere to lower depths than on Earth's opposite side. Astronauts orbiting Earth refer to this lowering as the South Atlantic Anomaly (SAA).

The anomaly is of concern because orbiting spacecraft carry astronauts through this area. Crews are exposed to higher concentrations of dangerous ions. Space crews may need additional radiation protection for those passages. It is not a good idea to be doing a space walk while passing through the SAA.

HOMELESS PIGEONS

Homing pigeons are remarkable birds. They have been known to find their way home across 1,000 miles (1,600 km) of unfamiliar territory. During World War I (1914–1918) and World War II (1939–1945), pigeons were used by the military to carry messages in tiny capsules attached to their legs.

People worldwide hold races for homing pigeons. Their remarkable navigation powers are based on their ability to sense Earth's magnetic field. It is as though the birds have an internal compass. These "compasses" work great unless the Sun erupts with major solar storms. During solar storms, the space weather in the outer atmosphere may become very intense. This causes magnetic field lines at low altitudes to fluctuate. The poor homing pigeons can't get a direction fix, so they end up lost.

WANDERING POLES

From about 121 to 83 million years ago, Earth's magnetic poles were very stable and didn't reverse once.

The wandering nature of Earth's magnetic poles can be observed by scientists. You can observe it too if you have a very good magnetic compass. Use the compass to find the direction of magnetic north. Then place a fixed marker, where it won't be disturbed, showing that exact direction.

In about ten years, go back to your marker and take a new measurement of magnetic north. Its direction will be different by a degree or two. Okay, coming back in ten years seems unrealistic. But scientists have been making observations of the direction of magnetic north from London, England, for hundreds of years. From 1580 to 1819, the north magnetic pole shifted a total of 36 degrees.

Another way to plot the movement of Earth's magnetic poles is to measure magnetic minerals in volcanic rock. When lava cools, mineral crystals form. Some of the crystals have magnetic properties. Like iron filings sprinkled around a magnet, the rock crystals align themselves with Earth's magnetic-force lines and are frozen in that direction within the rock. (Magnetic minerals also align themselves as sediments accumulate on lake and ocean bottoms. Over time, these minerals change into sedimentary rock.)

In this photomicrograph of volcanic rock in Oregon, the bright-colored mineral is magnetite. The magnetite within the rock would have aligned itself with the magnetic poles.

Scientists can map the orientation of magnetic minerals in volcanic rocks around the world. Using dating techniques to determine when the rocks formed, they can estimate the position of the magnetic poles at that time. Regardless of where the rocks are, the magnetic minerals in rocks of the same age point to the same place. However, magnetic minerals in rocks of different ages point in different directions. Looking at the directions of the minerals from rocks of different ages shows where the magnetic poles have been throughout Earth's history.

To make things even more complicated, the magnetic poles occasionally weaken and reverse themselves. Realignments in the circulation of the core cause the north and south magnetic poles to change places. North becomes south and south becomes north. The realignment is sort of like taking a bar magnet, standing it on its end so that north is up and south is down, then turning it upside down. Earth itself doesn't turn over, just its magnetic field. The reversals take place, on the average, every quarter of a million years. But the interval can range from a few tens of thousands of years to a few million years.

During the twentieth century, the magnetic north pole moved an average of 6 miles (10 km) per year. It has picked up speed to 24 miles (39 km) per year and, at its current rate, should move from Canada to Siberia in the next few decades.

For a time, during the reversal, the magnetic north and south poles weaken and disappear. Somehow, Earth's dynamo shuts itself off and then turns itself back on. Scientists determined this when they found rocks where the magnetic minerals were not aligned in any particular direction. The age of these rocks told scientists that the magnetic poles reversed 780,000 years ago. Why this happened is not well understood. Unfortunately, no scientists were around last time to study the reversal and discover why it happened.

Pole reversal is of great concern because of its potential effects on life on Earth. While Earth's dynamo is hard at work, the magnetic field lines emerge from Earth's crust and extend far out into space. When Earth's dynamo shuts down for a pole reversal, the magnetic field lines extending out into space disappear. Earth's primary protection against ions from the Sun and other stars disappear until the geomagnetic field reestablishes itself. Without the magnetic field, dangerous radiation can reach Earth's surface. Depending upon how strong the radiation is and how long the exposure lasts, the radiation could sicken or kill many life-forms.

Humans would be especially susceptible to the radiation. Eyes would be damaged, and skin would develop radiation burns. The incidence of cancer and leukemia would rise rapidly. Radiation protection technology could prevent much of the damage to humans, but sunbathing would definitely be prohibited. New products in the store might have labels stating "SPF-500!"

The location of Earth's north magnetic pole was first discovered in 1831. When Norwegian explorer Roald Amundsen reached the pole in 1904, its location had shifted 330 miles (530 km).

CHAPTER 4

Shields Up!

In spite of its next-to-nothingness, Earth's outer atmosphere is an amazingly complex place. Much of its complexity comes from Earth's magnetosphere. It interacts with the solar wind and deflects or captures the ions that make up the wind. In turn, the magnetosphere is shaped by the solar wind.

Without the solar wind, Earth's magnetic field lines would be symmetrical and reach out equally in all directions. Instead, the pressure of the wind distorts the field. It squashes the lines together on the side of Earth nearest the Sun while stretching them out on the opposite side. The effect is similar in appearance to the ripples of water in a river as the water passes around a projecting rock. The ripples are

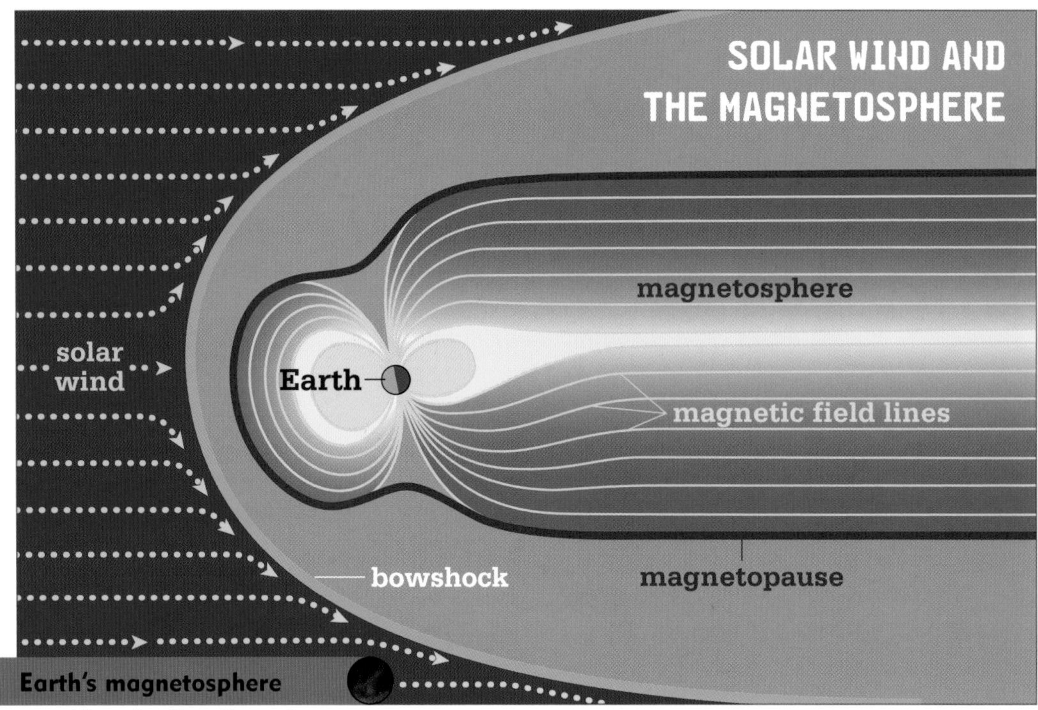

Earth's magnetosphere provides the first line of defense against the charged partical radiation in the solar wind.

close together on the upstream side of the rock. They stretch out in a teardrop shape on the downstream side.

The solar wind is able to shape the magnetosphere because of electromagnetism. We have learned that the particles that make up the solar wind are electrically charged. We have also learned that Earth's magnetic field lines result from electric currents within Earth. The magnetosphere's field lines attract the solar wind particles. Some of the particles are captured by the attraction of the field. Others are deflected to the sides by previously captured particles so that they pass around Earth. (This is similar to the way the north ends of two magnets repel each other. Two particles with the same charge, such as two positively charged protons or two negatively charged

electrons, repel each other.)

Because the ions are traveling at 1 million miles (1.6 million km) or more per hour, they transfer some of their momentum to the field lines when they are captured or deflected. Think of catching a thrown baseball. When you make the catch, your hands are pushed backward by the momentum of the ball. Even if the ball just glances off your hands, your hands have absorbed some of the ball's momentum. As the side of the magnetosphere facing the Sun catches or deflects the solar-wind particles, the field lines are pushed back.

The boundary between the solar wind and the magnetosphere is called the bow shock. The name comes from a similar distortion of water around the bow of a moving ship.

In this 2002 image from NASA's Hubble Space Telescope, a bow shock around a very young star is shown. The bow shock occurs when two streams of gas collide. In this case, the star emits a vigorous wind that hits the gas moving away from the center of a nearby nebula.

Depending upon how strong the solar wind is, the bow shock extends outward from Earth about 37,000 to 60,000 miles (60,000 to 100,000 km). On the side of Earth away from the Sun, the magnetosphere stretches out into space, like the wake of a ship, hundreds of thousands of miles.

Variations in the size of the magnetosphere are due to activity on the Sun. During quiet times, the front end of the magnetosphere stretches farther toward the Sun because the solar wind is weaker. During stormy periods on the Sun, the solar wind strengthens. This compresses the field lines and increases the number of particles captured by the field lines.

VAN ALLEN RADIATION BELTS

On January 31, 1958, the first U.S. artificial satellite was launched. The satellite, *Explorer 1*, was a 30-pound (14 kg), pencil-shaped vehicle.

***Explorer 1* launched from Cape Canaveral, Florida, on January 31, 1958, four months after the Russian satellite *Sputnik 1* was launched.**

DUCK!

There is a good reason why the outer layers of space suits are made with fabrics that can stop a bullet. The outer atmosphere is a big shooting gallery of tiny particles traveling many times the speed of the fastest bullets. Most of the particles come from deep space. They are tiny grains of dust from comets and sand-grain-sized rocks and metal flakes broken off during collisions of big space rocks such as asteroids.

Another source of particles is space pollution. Little paint chips and metal filings are shed by rockets and satellites. Fragments from the occasional vehicle explosion in orbit have also peppered the outer atmosphere.

Earth's gravity draws most particles into the lower atmosphere. Friction with denser and denser air causes them to burn up. Other particles go into orbit around Earth for a time. Collisions with gas molecules eventually bring them down.

Astronauts need tough space suits because of the speed of the particles, which travel 10, 20, 40 or more miles (16, 32, 64 or more km) per second. Though the particles are tiny and don't have much mass, their speed causes them to pack a real wallop. They carry enough energy to blast out small craters in skin.

U.S. astronaut James F. Reilly wears his space suit as he participates in a space walk at the International Space Station in July 2001. Astronauts' space suits are made to withstand the fast-moving particles in outer space.

It had four flexible whip antennae protruding from its sides. The satellite carried several instruments. One of which was a Geiger counter (radiation detector) designed to study cosmic rays. It made the first measurements of ion radiation around Earth.

The Geiger counter performed well at low altitudes, but it mysteriously recorded zero radiation counts at higher levels. Scientists had expected Earth's magnetic field to trap ions, causing low counts at low elevations and high counts at high elevations. The *Explorer 1* data seemed to show just the opposite. A few months later, the *Explorer 3* satellite sent data that resolved the issue.

Not all charged particles in the solar wind are deflected by Earth's magnetic field. Some particles are caught by the field and are concentrated in two or more radiation belts that surround Earth's middle regions.

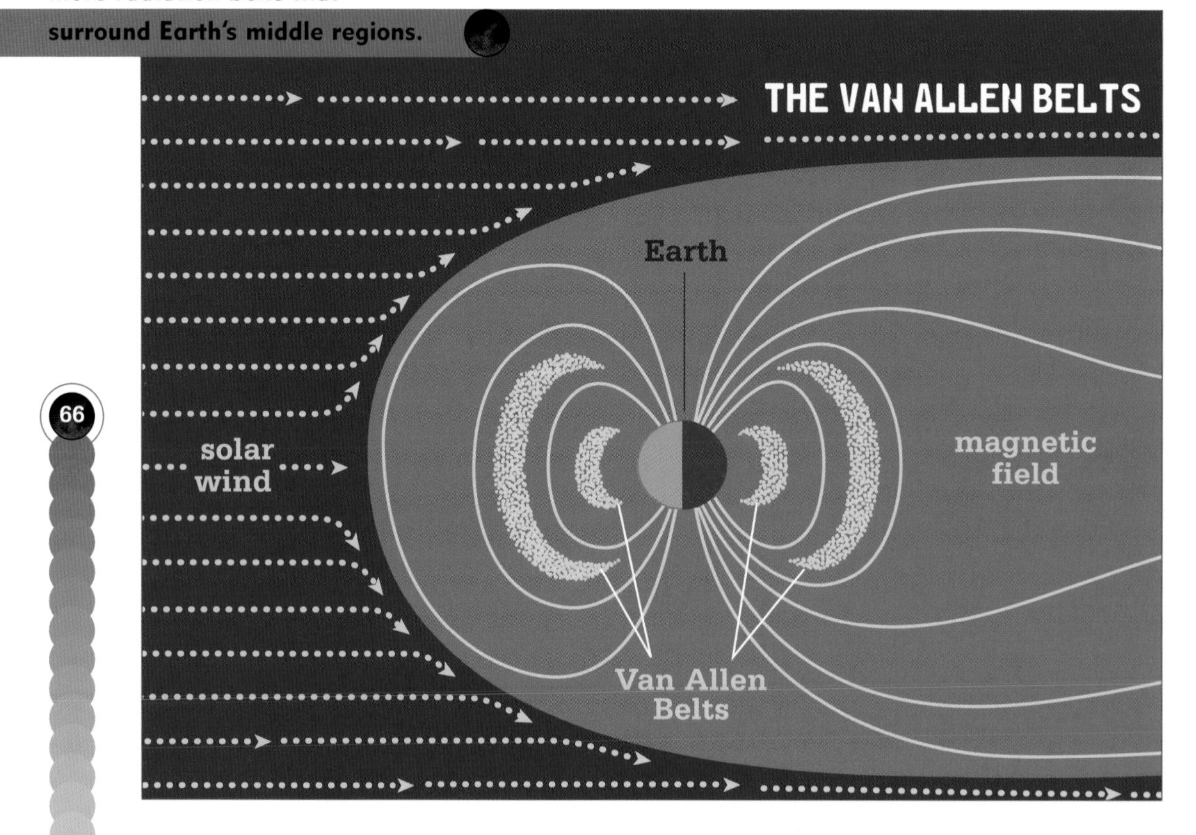

Explorer 1's radiation counts were in error. So many energetic particles were in the magnetosphere that the Geiger counter was overwhelmed. It took an electronic version of a coffee break. Measurements taken during Explorer missions and by later satellites helped scientists discover two doughnut-shaped belts of dangerous concentrations of ions surrounding Earth. The belts were named after James Van Allen, who led the University of Iowa team that designed and built *Explorers 1* and *3*.

The Van Allen belts consist of charged solar-wind particles trapped by Earth's magnetic field. The inner belt stretches from about 400 miles (640 km) to approximately 3,915 miles (6,300 km) above the surface. The greatest intensity in the inner belt is usually found between 1,243 and 3,110 miles (2,000 and 5,000 km) from Earth's surface.

WHO WAS JAMES VAN ALLEN?

James A. Van Allen was born in Iowa in 1914. He became a scientist in the field of terrestrial magnetism and worked on developing instruments for V-2 rockets. Van Allen led the team that developed the first sounding rocket, the Aerobee. Sounding rockets are designed to take measurements of the upper atmosphere and near space. In the late 1950s, Van Allen developed the radiation detectors that flew on the first and third Earth satellites. The detectors discovered the radiation belts surrounding Earth.

The outer Van Allen belt starts at about 6,200 miles (10,000 km) above the surface and extends to about 40,000 miles (64,000 km). Its greatest intensity is found between 9,010 and 11,810 miles (14,500 to 19,000 km). During intense periods of solar activity, the ranges of the belts are modified by increased solar particle pressure. Their levels of radiation also increase.

The Van Allen radiation belts provide a kind of radiation armor that astronauts depend upon when orbiting Earth. As long as a space mission remains within the inner belt, radiation exposure is moderate. During unusually strong solar activity, the danger increases. Then crews on the International Space Station stay in areas of the station that have greater shielding.

The Van Allen belt radiation armor is not without its chinks. In places, radiation is able to penetrate more deeply toward Earth's surface. One such area is the South Atlantic Anomaly near the coast of South America.

CONCLUSION

Poker Flat, Alaska

A 73-foot-long (22 m) Black Brant rocket rests on a steel launch support gantry. It is about 4 A.M., the air is still, and the temperature is –31°F (–35°C). The small launch team has been waiting for the right moment. Overhead, the sky is alive with rippling auroral colors. The team waits for a satellite to pass overhead. The idea is to launch the rocket when the aurora is strong, the upper-atmosphere winds are mild, and the satellite is above. It is a particular combination that has taken months to come together.

Auroras are one of the most beautiful and mysterious phenomena of Earth's outer atmosphere. If you live in the far north or far south regions of Earth, you are probably very familiar with

Aurora australis, or the southern lights, as seen from South Australia. The colors in auroral rays indicate what gases are glowing in the outer atmosphere.

them. Auroras take on many shapes. They swirl, flicker, or move across the sky like curtains and rays.

Auroral rays follow the direction of the invisible field lines of Earth's magnetosphere. They are best seen at night during the long winters, when the sky gets very dark for many hours. Curtains or rays of greenish, reddish, or yellowish light move across the sky. These colors have meaning. They tell us what gases are glowing in the outer atmosphere.

Gases in the outer atmosphere glow when their atoms are struck by solar particles. The electrically charged atomic particles are captured by Earth's magnetosphere. They spiral around the field lines as they travel back and forth between the magnetic poles. Where the field lines dip into the atmosphere, the particles strike gas atoms. When struck, the atoms release photons of light in particular wavelengths (colors).

Some elements, such as oxygen, can give off a range of different colors (pale green, yellowish green, and even red). The color depends upon how high above Earth the atoms are when they are struck. Nitrogen gas can give off a violet light, which is really hard to see. Sometimes it also glows red. Hydrogen and helium give off violet or blue light, but it usually takes a camera with sensitive film to detect these colors. During strong periods of solar activity, auroras

LISTENING TO THE LIGHTS

Many people of the far north claim to hear crackling or faint whistling sounds during strong displays of the aurora. Scientists are not sure where the sounds come from. During auroras, charged particles flow to Earth along magnetic field lines and induce electric currents in the ground. The crackles and whistles could be the sounds of electric discharges from tree branches.

Regardless of their cause, some native peoples believe that if you hear the whistling sound, you only have to whistle back and the lights will come closer to you. They believe the lights are spirits of departed people who play an endless football-like game in the sky.

may have different colors because atoms at different altitudes are also glowing.

By 4:57 A.M., everything is ready. Traffic on the nearby Steese Highway is stopped. Airplane traffic has been routed away from the area. The count reaches zero. In a flash and a roar, the rocket leaves a trail of smoke as it streaks skyward. In moments a great explosion occurs in the heart of the aurora. A cloud of barium vapor is dispersed. At first the cloud is yellowish green and about the size of the full Moon. In just seconds, the cloud changes to reddish purple as the

THE DARK OF THE NIGHT

Have you ever wondered how it is possible to see objects on a dark night? Not taking a walk down a street with the lights from homes and streetlights. Instead, in the country, far from electrical lights. If the Moon is out, seeing is easy. You may actually be able to read a newspaper when the Moon is full.

What about when the Moon isn't up? You won't be able to read the newspaper, but you can still avoid falling in holes and stumbling over stumps. How can you see? It's not starlight that is helping you but a phenomenon called zodiacal light. The faint, diffuse background light coming from the sky above is actually sunlight that is reflected off interplanetary dust!

barium particles are ionized by sunlight. Soon the cloud begins changing shape.

The purpose of the experiment is to learn about the shape of Earth's magnetic field. Observers on the ground and in the satellite above track the movements as the cloud evolves into a long series of rod shapes that are aligned around the magnetic field lines. Just eight minutes after liftoff, the rocket has completed its mission and crashes on the ice pack covering the Arctic Ocean.

The University of Alaska operates the Poker Flat Research range. The nearly 10-square-mile (26 sq. km) facility is about 30 miles (50 km) north of Fairbanks. In winter the Sun is so low to the horizon that pitch-black nights last twenty hours. It is a perfect place to study the aurora borealis.

The small Poker Flat team of scientists and technicians belong to a large family of scientists who dedicate themselves to understanding planet Earth. They are probing Earth's core with sound waves, riding to the ocean's depths in submersibles, collecting rock samples from mountain peaks, measuring the winds, investigating bacterial slime in deep caves, and lofting instruments to the edge of space. This team is piecing together the story of Earth.

While each team focuses on a particular aspect of Earth, they all know that none of Earth's spheres exist and function by themselves. The core and mantle, the lithosphere, hydrosphere, atmosphere, biosphere, and the outer atmosphere are parts of the whole. Together, they are planet Earth.

GLOSSARY

atom: the smallest part of a chemical element that has all the properties of that element. Atoms consist of a nucleus of protons and neutrons surrounded by orbiting electrons.

aurora: colorful displays of light caused by the interaction of charged particles with Earth's magnetic field in the upper atmosphere

bow shock: the compression of the field lines of Earth's magnetosphere by the pressure of charged particles emitted by the Sun

CME (coronal mass ejections): a huge, bubblelike storm that erupts from the Sun's atmosphere and showers the solar system with charged particles

convection currents: vertical motions caused by the rise of hot fluids and the fall of cool fluids

corona: the atmosphere of the Sun

Curie point: the temperature above which a magnetic material loses its magnetic field

dynamo effect: the process within Earth's outer core that produces the planet's magnetic field

electromagnetic spectrum: the entire range of wavelengths of radiation including radio waves, infrared light, visible light, ultraviolet light, X-rays, and gamma rays

electromagnetism: magnetism produced by electric currents

electron: a negatively charged subatomic particle. Electrons orbit the nucleus of an atom.

exosphere: one of the outermost regions of Earth's atmosphere

flare: a violent explosion from the Sun's surface

fusion: the nuclear process, taking place in the Sun's core, that converts hydrogen to helium and energy

gamma rays: the most energetic form of electromagnetic radiation

geomagnetic field: Earth's magnetic field

heliopause: the zone where the solar wind gives way to interstellar space

ionosphere: an outer zone of Earth's atmosphere where gas atoms are ionized and become electrically charged

ions: electrically charged atoms or groups of atoms

magnetic field: a region in which magnetic forces can be observed

magnetosphere: the space around a star or planet where its magnetic field can be detected

mantle: the thick layer of rock between Earth's core and the crust

mesosphere: the zone of the atmosphere that begins 50 miles (80 km) above Earth's surface, where thin gases are electrically charged due to energy absorbed from the Sun

meteor: an object that enters Earth's atmosphere and is heated by collisions with air molecules

molecule: two or more atoms held together by chemical bonds. A molecule is the smallest unit of a chemical compound that has all the properties of that compound.

neutron: a subatomic particle with no charge. Neutrons are found in the nucleus of an atom.

neutron star: a star consisting mostly of neutrons

ozone layer: a thin layer of ozone gas within Earth's stratosphere that protects Earth's surface from dangerous ultraviolet radiation

payload: the satellite, space probe, or astronaut crew carried by a rocket

photon: a particle of light

plasma: a state of matter consisting of ionized gas

proton: a positively charged subatomic particle. Protons are found in the nucleus of an atom.

radiation: energy that is transmitted in the form of rays, waves, or particles

solar wind: an outward flow of charged particles (electrons and protons) coming from the Sun

stratosphere: the middle, quiet layer of Earth's atmosphere where jet aircraft fly

sunspot: a magnetic storm on the Sun's surface that looks dark because its temperature is lower than that of the surrounding surface

thermosphere: an upper region of Earth's atmosphere where temperature increases with altitude

troposphere: the lowest layer of Earth's atmosphere, where most of Earth's weather takes place

BIBLIOGRAPHY

Carlowicz, Michael, and Ramon Lopez. *Storms from the Sun: The Emerging Science of Space Weather*. 2nd ed. Washington, DC: National Academies Press, 2002.

Freeman, John W. *Storms in Space*. Cambridge, UK: Cambridge University Press, 2001.

Hamblin, W. Kenneth, and Eric H. Christiansen. *Earth's Dynamic Systems*. 10th ed. Upper Saddle River, NJ: Prentice Hall, 2004.

Lutgens, Frederick K., Edward J. Tarbuck, and Dennis Tasa. *The Atmosphere: An Introduction to Meteorology*. 10th ed. Upper Saddle River, NJ: Prentice Hall, 2006.

Rozelot, Jean-Pierre, ed. *Solar and Heliospheric Origins of Space Weather Phenomena*. Lecture Notes in Physics series. New York: Springer, 2006.

Wallace, John M., and Peter V. Hobbs. *Atmospheric Science: An Introductory Survey*. 2nd ed. Vol. 92. San Diego: Academic Press, 2006.

FOR FURTHER INFORMATION

Books

DK Publishing. *Space Exploration*. New York: DK Children, 2004.

Hall, C. *Northern Lights: The Science, Myth, and Wonder of Aurora*. Seattle: Sasquatch Books, 2001.

Harrison, C., and D. Krasnow. *Weather and Climate*. Discovery Channel School Science series. Milwaukee: Gareth Stevens Publishing, 2004.

Miller, Ron. *Earth and the Moon*. Minneapolis: Twenty-First Century Books, 2003.

———. *The Sun*. Minneapolis: Twenty-First Century Books, 2002.

Savage, C. *Aurora*. Ontario: Firefly Books, 2001.

Scholastic. *Scholastic Atlas of Weather*. New York: Scholastic Reference, 2005.

Websites

The following Internet sites are good places to look for more information and pictures about the outer atmosphere and magnetosphere:

The Aurora Page
http://www.geo.mtu.edu/weather/aurora/
This website provides links and images about the northern lights.

Earth's Atmosphere
http://liftoff.msfc.nasa.gov/academy/space/atmosphere.html
Learn about the layers of the Earth's atmosphere as well as its compositional make up.

Earth's Magnetic Field
http://liftoff.msfc.nasa.gov/academy/space/mag_field.html
Explore specific information about Earth's magnetic field.

The Exploration of the Earth's Magnetosphere
http://www-istp.gsfc.nasa.gov/Education/Intro.html
This detailed website provides an overview of research that was done on Earth's environment.

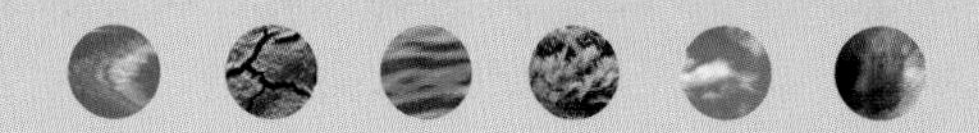

National Geophysical Data Center (NGDC)—Magnetic Declination
http://www.ngdc.noaa.gov/seg/geomag/declination.shtml
This website, compiled by the National Geophysical Data Center, provides information on magnetic declination. Links offer activities to compute declination as well as maps that allow you to explore different types of declinations.

Poker Flat Research Range
http://www.pfrr.alaska.edu/pfrr/index.html
Visitors to this site will find information on the Poker Flat Research Range. Links provide overviews of research that was conducted on Earth's auroral and middle to upper atmosphere.

The Solar and Heliospheric Observatory
http://sohowww.nascom.nasa.gov/
This website provides information on the goals and status of the Solar and Heliospheric Observatory (SOHO) mission.

Space Weather Center
http://www.spaceweathercenter.org/
This website provides information about the sun, Earth's magnetic field, magnetic storms, the weather in space and how space is researched.

Today's Space Weather
http://www.sec.noaa.gov/today.html
This website is ideal for students and teachers interested in space. It provides in-depth information, including graphs, of the current weather in space.

INDEX

ABOUT THE AUTHOR

Gregory L. Vogt holds a doctor of education degree in curriculum and instruction from Oklahoma State University. He began his professional career as a science teacher. He later joined NASA's education programs teaching students and teachers about space exploration. He works in outreach programs for the Kennedy Space Center. He also serves as an educational consultant to Delaware North Parks Services of Spaceport and is the principal investigator for an educational grant with the National Space Biomedical Resesearch Institute. Vogt has written more than seventy children's science trade books.

PHOTO ACKNOWLEDGMENTS

The images in this book are used with the permission of: PhotoDisc Royalty Free by Getty Images, (lava: center ring), (cracked earth: second ring), (vegetation: fourth ring), (sky/clouds: fifth ring), all backgrounds, pp. 2–3; MedioImages Royalty Free by Getty Images, (water: third ring), all backgrounds, p. 2; NASA (stars/nebula: main ring and sixth ring), (earth), all backgrounds, pp. 2–3, 6, 7, 9, 11, 12, 13, 22, 30, 44, 46, 57, 59, 60, 61, 69; © Laura Westlund/Independent Picture Service, pp. 10, 16, 19, 21, 32, 33, 38, 40, 41, 42, 47, 51, 52, 55, 62, 66; © Ali Jarekji/Reuters/ CORBIS, p. 27; NASA Marshall Space Flight Center (NASA-MSFC), pp. 29, 64, 65; © Getty Images, pp. 35, 45, 70; © Hinrich Baesemann/epa/CORBIS, p. 43; © Gregory L. Vogt, p. 49; © Tony Hutchings/Stone/Getty Images, p. 54; U.S. Geological Survey/photo by Richard Reynolds, p. 58; NASA Goddard Space Flight Center (NASA-GSFC), pp. 63. Front Cover: PhotoDisc Royalty Free by Getty Images, (cracked earth: second ring), (lava: center ring), (vegetation: fourth ring), (sky/clouds: fifth ring); MedioImages Royalty Free by Getty Images, (water: third ring); NASA, (stars/nebula: main and sixth ring). Back Cover: NASA, (stars/nebula), (earth). Spine: NASA, (stars/nebula).